Psychology as a Moral Science

Svend Brinkmann

Psychology as a Moral Science

Perspectives on Normativity

Springer

Svend Brinkmann
Department of Communication and Psychology
University of Aalborg
Kroghstræde 3
9220 Aalborg
Denmark
svendb@hum.aau.dk

ISBN 978-1-4899-9014-3 ISBN 978-1-4899-9014-3
DOI 10.1007/978-1-4419-7067-1
Springer New York Dordrecht Heidelberg London

Printed on acid-free paper

Springer is part of Springer Science+Business Media (www.springer.com)

For Signe

Foreword

Since the 1970s, a steadily mounting wave of criticism has threatened to engulf what passed as psychology in the mid-twentieth century. The grounds of complaint have ranged from the irrelevance of "laboratory" psychology to any issue of everyday life, to fundamental objections to the conceptual naiveté of academic psychology, in particular the uncritical adoption of a causal metaphysics as the structuring principle of the flow of human thought, action, feeling, and perception. Among the sources of the pseudoscientific nature of mainstream experimental psychology has been a prevailing ignorance of the natural sciences adopted as ideals, and a steadfast refusal to take account of the role of moral orders in the formation and management of human life forms.

The effect of 50 years of efforts at reform can now be seen in the growth of qualitative and cultural psychologies as significant components of a well-rounded and useful training in the basic elements of genuinely scientific psychology. It is scarcely credible that even a decade or two ago students could be introduced to the principles of social psychology without the central role of language as the medium of social interaction even being mentioned! Choice of pronoun can have profound consequences for a social relationship if you are French or Japanese. In this and many other psychologically relevant matters, the overwhelmingly Anglophone character of psychology has stood in the way of forging an authentic identity for psychology as a discipline. These developments should have brought the tacit subscription to a causal metaphysics under scrutiny and stimulated reflection on the ultimate consequences of tasking up the insight that psychological phenomena are meanings and that the principles of their ordering into coherent processes are normative. Indeed they have to some extent. However, the most profound consequence that is the focus of this study is the renewed emphasis on psychology as a moral science, much in the way that the nineteenth-century pioneers of a scientific approach to understanding personal and social life was proposed.

In this important book, Svend Brinkmann has provided a brilliantly argued and, one hopes, a definitive account of how psychology will look when the shift to an explicit moral science has finally come about. The thrust of the argument is to show that moral issues and concepts as to how one should live as a human being among others are not just add-on bits to the psychological paradigms that already exist, but are the very roots from which psychology should spring. We do not need a psychology of morality, in the manner of Piaget or Kohlberg, but a moral science. Emotions are not to be taken as displays of this or that moral judgment, but as the products of the moral orders in which human beings live. Psychology should self-consciously return to the "ethology" in the original acceptation of the word, the study of meaningful conduct.

This is the "interpretative-pragmatic" view of what psychology ought to become. A key step in the overall project is the setting aside of the famous Humean claim that factual and valuational aspects of the discourses that make up the substance of human life are radically disjoint – the refusal to accept the alleged naturalistic fallacy. The argument that defuses this famous "fallacy" is as simple as it is profound. All accounts of morality must rest on evaluative premises – true – but aren't they independent of the facts of human life, such as those the evolutionary psychologists reveal with their hypothesis that the lineaments of our present lives were laid down in the conditions of the paleolithic era? But that life was itself ordered in accordance with norms – how else could it have been? Our hominid ancestors did not simply respond in ways that their genes predisposed them to. Our genetic endowment is the result of the normative framing of life along the banks of the lakes in the Olduvai Gorge. In a more philosophical vein, we have the arguments of Wittgenstein and Searle to the effect that the very possibility of a rule-ordered life depends on the existence of public institutions – morality cannot be based on the accidental or even forced coherence of private, subjective feelings of pleasure or approval. Neither can one obey a rule only once nor can one sustain a normative framework for action alone – how could one know that one had remembered yesterday's rule correctly if one had only one's own memory as an authority?

One can only hope that this subtle and profound analysis of the proper foundations for a science of people thinking, acting, feeling, and perceiving will attract the attention it deserves. Finally, this may be the last push that enables psychology to turn the corner from the darkness of conceptual confusion into the light of a moral science.

Georgetown
April 2010

Rom Harré

Preface

The ideas of this book were originally presented in my PhD dissertation, which I defended in 2006 at the University of Aarhus, Denmark. The text has since been thoroughly revised, new chapters have been added, and old ones deleted. The contents of some of the chapters have also appeared as individual journal articles, but I have tried in this book to state my arguments in the form of a coherent, book-length account of psychology's complex relationships to moral issues.

This text contains materials that have previously appeared in:

Brinkmann, S. (2009). Facts, values, and the naturalistic fallacy in psychology. *New Ideas in Psychology, 27(1)*, 1–17.

Brinkmann, S. (2008). Changing psychologies in the transition from industrial society to consumer society. *History of the Human Sciences, 21(2)*, 85–110.

Brinkmann, S. (2006). Mental life in the space of reasons. *Journal for the Theory of Social Behaviour, 36(1)*, 1–16.

Brinkmann, S. (2005). Human kinds and looping effects in psychology: Foucauldian and hermeneutic perspectives. *Theory & Psychology, 15(6)*, 769–791.

Acknowledgments

I wish to express my gratitude to my PhD supervisor, the late Steinar Kvale, from whom I learned how to be an academic. I am grateful in so many ways for what I learned from working with and for Steinar. I would also like to thank the members of my PhD committee, Rom Harré, Alan Costall, and Benny Karpatschof. They provided me with valuable feedback, and I have learned a lot from conversations with all of them (and from reading their work). I am particularly thankful to Rom Harré, who encouraged me to publish my ideas as a book. As a PhD student I profited greatly from discussions with Shyam Cawasjee, Donald Polkinghorne, Lene Tanggaard, Claus Elmholdt, Peter Musaeus, Jacob Klitmøller, Johan Trettvik, and Klaus Nielsen. I am particularly grateful to Daniel N. Robinson for his proficient supervision and valuable discussions, especially during my visits to Oxford University in 2002 and 2004, where many of the ideas of this book were conceived. I would also like to thank my new colleagues at the University of Aalborg, where I have worked while I have finished the present text. I am thankful to everyone for welcoming me in a very kind way. Finally, I want to thank my wife Signe Winther Brinkmann for her encouragement and acceptance of the strange practice of academic work. I dedicate this book to her, *sine qua non*.

Contents

About the Author

Svend Brinkmann is Professor of Psychology in the Department of Communication and Psychology at the University of Aalborg, Denmark. His research areas are general psychology and qualitative methods, and he is codirector of the Center for Qualitative Studies at the University of Aalborg and also editor of the journal *Qualitative Studies*. Svend Brinkmann has published books in Danish about the mind, identity, and the philosopher and psychologist John Dewey, and he is coauthor (with Steinar Kvale) of the English language book *InterViews*. In addition, he has published several journal articles about the philosophy of psychology, qualitative methods, moral inquiry, and approaches to human science such as pragmatism, hermeneutics, and discourse analysis.

Chapter 1
Psychology and Morality: An Interpretive-Pragmatic View

This book is about psychology's grounding in morality – or, in other words, about the ethical foundations and implications of psychology.[1] It presents the argument that psychological phenomena are inherently moral phenomena, and that psychology, as an array of investigative and interventionist practices, is, and ought to be, a moral science. Throughout the book, I aim to present a unified view of psychology and morality, not as two disjointed fields that are accidentally brought together, but as deeply and inherently related in many different ways. Often, however, the relations between psychology and morality are not recognized by psychologists themselves and this, I argue, is detrimental to the discipline, but also to the society that is affected by the workings of psychology in many different ways. Part I begins with a number of critical investigations into how modern psychology has shaped and in some ways distorted our views of morality and ourselves, and part II advances more positive and prescriptive views about how properly to conceive of morality and its relation to psychology.

What this book aims to say can be summed up in two theses: the first is that psychological phenomena are normative, and the second is that not all normativity is conventional. Morality is one important kind of non-conventional normativity (but there are others such as the normativity of logic). Psychology is essentially a contested and fragmented discipline with numerous interests in how the discipline should be designed. My suggestion in what follows is that psychology ought to look more like moral philosophy than anything else. This may at first sight strike the reader as deeply misguided, but I hope to provide evidence that this suggestion is at least as reasonable as other current arguments that it should look more like neurochemistry (cf. certain neuroscientists), computer science (cf. some cognitive scientists), biology (cf. evolutionary psychologists), sociology (cf. social constructionists), anthropology (cf. cultural psychologists), literary studies (cf. postmodernists), or

[1] In general, I use the terms "morality" and "ethics" interchangeably. "Ethics" comes from the Greek *ethos* (character) and "morality" from the Latin *mores* (which also means character, custom, or habit) (Annas, 2001). I work with a broad definition of these concepts to refer to *the oughtness of human existence*, i.e., to the idea that human life is not just one factual state of affair after another, but centrally involves non-arbitrary and non-conventional normative demands (to act, think, feel, and be in required ways).

S. Brinkmann, *Psychology as a Moral Science: Perspectives on Normativity*,
DOI 10.1007/978-1-4419-7067-1_1, © Springer Science+Business Media, LLC 2011

political science (cf. critical psychologists). I am aware that my suggestion is radically different from the so-called mainstream psychology, but I hope that skeptical readers of this book will occasionally remind themselves that other existing suggestions, like those I just mentioned, are at least as radical – although people have grown used to these suggestions, which means that we have lost sight of just how radical – and radically wrong – they are (e.g., the attempts to design psychology as neurochemistry or computer science).

I aim to present and develop a coherent outlook on psychology and morality. I call this outlook "interpretive-pragmatic."[2] Concerning moral psychology – the field that studies the psychological dimensions of morality – the term "interpretive" should point toward my belief that moral judgment in practice necessarily involves situational interpretation and judgment, and cannot be understood as a mere application of moral rules, and "pragmatic" should emphasize the idea that the validity of moral judgments, rules, and concepts is to be evaluated by their effects in practical action. In an interpretive-pragmatic approach to psychology more broadly conceived, "interpretive" indicates that psychological phenomena are situated in concrete contexts that demand situational interpretation in order to be understood, and "pragmatic" indicates that psychological phenomena are first and foremost to be thought of as aspects of our practical dealings with the world and each other (Polkinghorne, 2000). Being alive as a human being is, from the interpretive-pragmatic perspective, an interpretive process of inquiry that is subject to normative appraisal. Psychological phenomena and moral phenomena are, in a broad sense, conversational (Harré, 2004). The interpretive-pragmatic view on morality will be contrasted with various forms of moral theory that exist in contemporary psychology. On the one hand, emotivist, subjectivist, and relativist theories are discussed and ultimately rejected (e.g., socio-biological and social constructionist theories of morality), and, on the other hand, various kinds of formalism and proceduralism are shown to be inadequate (e.g., neo-Kantian theories). The interpretive-pragmatic view steers a middle course between emotivism and proceduralism, i.e., between the view that morality is nothing but a matter of subjective emotional likings and the view that it is ultimately derived from some universal procedure.

I deliberately refer to the interpretive-pragmatic view as a "view" or an "outlook," rather than a "theory." For unlike sociobiologists or Kantian psychologists like Lawrence Kohlberg, I do not intend to develop anything that can properly be called a "theory" about morals. Rather, my aim is to direct psychologists' attention to certain moral aspects of human experience that we simply cannot ignore or eliminate, if we want to remain true to the phenomena that psychologists deal with. The present study is thus committed to a phenomenological outlook on moral and psychological

[2] The interpretive-pragmatic view owes much to John Dewey's pragmatism, to Charles Taylor's hermeneutic theory of values, action, and identity, and to Aristotle's virtue ethics (especially as it has been developed by Alasdair MacIntyre). In addition, the broad field of discursive psychology, especially in Rom Harré's version, has been inspirational. Like Dewey's pragmatism and Taylor's hermeneuticism, Harré claims that the conversation is the most useful model for understanding psychological phenomena. As unfolding episodes, conversations are structured, not by causal laws, but by normative demands that should be studied by psychology.

inquiry that goes back to Aristotle (see Nussbaum, 1986), and was continued in John Dewey's naturalistic empiricism (Dewey, 1925). Dewey argued for the primacy of human experience and "the ordinary qualitative world" (Dewey, 1929:103). He understood theories not as mirrors of an independent reality, but as tools, whose function is to assist humans in solving problems and improving the ordinary qualitative world. Dewey claimed more specifically that it is a grave mistake to think about moral theory as "something other than, or something beyond, an analysis of conduct" (Dewey, 1891:187). If we must talk about moral *theory*, we should simply see it as "the theory of practice" (p. 187). A similar view is defended in this book, where morality is analyzed not as something over and above human conduct, but as an aspect of all human conduct *qua* conduct. I also follow Dewey in arguing that psychological theories are valid to the extent that they enrich the human world of social practices. Validity will thus be presented as a moral issue in psychological theorizing, and morality itself is presented as the bedrock of psychological phenomena. Morality, therefore, cannot be explained in psychological terms. It is rather the other way around: We should explain psychology in moral terms.

I shall argue that if we set out to "explain" morality at all, we should be careful not to do so by using different (non-moral) terms, or by measuring it in different dimensions. In my view, no theory about morality can ever be as well established as the moral phenomena of the ordinary qualitative world that it sets out to explain. Qualitative moral events ought always to be the starting point for our theoretical investigations. In our everyday lives we are much more certain of the correctness of the belief that we ought to try and save a drowning child, for example, than of any theory we can invoke to back that belief.

The main conclusion of the book is that psychology and its phenomena (human perceiving, acting, feeling, thinking, learning, and development) cannot be understood if the moral normativity of the human world is ignored or left out. Psychological phenomena are saturated with morality. The point can be put even stronger: The scientific project of exorcizing morality from psychology leaves no genuine psychological phenomena behind, for psychological phenomena are constituted by normative moral orders. My claim is that moral normativity should be seen as a *precondition* for what we call psychological phenomena, and not as a resulting *epiphenomenon* of psychological operations[3] (the latter view will be called a *psychologization* of morality[4]). Morality is not like butter on psychological bread, but rather like the flour that goes into making any psychological bread in the first place. Implicitly or explicitly, all psychological theories are moral theories, and all psychological practices are moral practices. The positive recommendation in this book is for psychologists to make explicit and discuss

---

[3] A similar argument is found in the influential moral phenomenology of Emmanuel Lévinas and his followers in psychology (Williams & Beyers, 2001; Williams & Gantt, 1998, 2002).

[4] Morality is psychologized when conceived as the result of psychological reactions, operations, likings, or desires. As will be argued in this study, moral values and the reasons for action that they provide are not psychological properties of the agent (e.g., desires), but aspects of our world of human interaction and social practices (this point is argued philosophically in Dancy (2000)).

the moral values they live by, since this could advance the discipline – scientifically as well as ethically.

Depsychologizing Psychology

Initially, the postulated unity between psychology and morality may sound puzzling. What does it mean when I say that "psychological phenomena are constituted by normative moral orders?" Explaining this is the main task of the entire book, but a few words may be helpful at this stage. What it means is that whenever we talk about something as a psychological phenomenon, rather than, say, a physiological phenomenon, we talk about something that exists in normative rather than causal connections. We talk about something that exists in what philosophers like John McDowell (1998) call "the space of reasons." Wilfrid Sellars (1997) originally described the space of reasons as a logical space of "justifying and being able to justify what one says." (p. 76). It is not the space studied by the physical sciences, where entities are related to each other in causal connections. This is a non-normative "space of causation."[5] The difference between the space of reasons and the space of causation can be illustrated with the help of two simple examples (cf. Brinkmann, 2006b):

1. A doctor taps on my knee and my leg moves. This is a pure physiological or reflexive process that can be fully explained in causal terms. We can state the *cause* of the leg's movement, but we cannot meaningfully articulate a *reason* for its moving. The event is not as such a psychological or mental event. Of course, the doctor had a reason to tap on my knee (i.e., to examine my reflexes), but the movement of the leg as such is causally determined: *I* had no reason to move my leg.
2. I watch a sad movie and begin to cry. Initially, this looks very much like the first episode, because it appears that the movie is the cause of my sadness. But there is an important difference that makes this second event a psychological event. For in this case, I can state a *reason* for the change in my mood ("when the woman left him, it was so sad that I had to cry"). This is not the same as stating a cause, for in this second case I can reasonably be asked to *justify* my reaction. Was it warranted? Perhaps I did not understand the movie; perhaps it was full of irony and in fact carried some quite happy and optimistic messages to the onlooker. By becoming aware of this, I may be brought to understand that I had no *reason* to cry. The first episode differs from the second in this regard, because

[5] I am aware that the metaphor of "spaces" has the unfortunate consequence that one is easily led to think dualistically of two disjointed worlds (nature vs. culture, causation vs. normativity) that have nothing to do with one another. I am not wedded to this metaphor, and I am skeptical of its dualistic connotations, but I do believe that it serves as a useful way of pointing to the important differences between material and intentional (normative, moral, etc.) properties of the world, as long as we remember that they are properties of *a* world (and not two different worlds in themselves).

> in the first case we cannot meaningfully say that "I was wrong" (or right for that matter) in moving my leg. But the second episode implies a psychological phenomenon, since it is subject to normative and even moral evaluation. It could be that my reaction expressed a deep, moral sensitivity on my behalf (and thus was morally praiseworthy), or it could be that my reaction was "too much" and improper (and morally blameworthy). Such moral questions are relevant with regard to psychological phenomena.

Philosophers like McDowell, who work in the tradition of Wittgenstein, argue that psychological phenomena cannot be understood as existing in physical space. The questions "where is your thought?" or "where is my hope?" do not make sense if understood as asking for a physical localization. Instead, psychological phenomena exist in a normative space, the space of reasons: "mental life is lived in the space of reasons" (McDowell, 1998:296). The present study is based on the premise that what it means to have "mental life," or to have a "mind," is not to have inner, mental representations or processes (that are somehow mysteriously thought to be able to represent "the outside world"). Rather, to have a mind means to be able to respond to normative *meanings*, rather than to mere physical stimuli (Dewey, 1916:29). Dewey conceived of "mind" as "the body of organized meanings by means of which events of the present have significance for us" (Dewey, 1934:273). The mind is not a thing that can be localized (e.g., in the brain), but rather an array of skills and dispositions that can be evaluated according to norms and standards most of which are local, although some, I shall argue, are universal. Some of the most important universal norms are what is conventionally referred to as moral values (e.g., truthfulness), and these should rightly be seen as foundational for complex forms of human psychological life.

As creatures with a mind, human beings are creatures who can respond to meanings, and have been taught "what is a reason for what" (Lovibond, 2002:21; McDowell, 1994:126). They are creatures who have been taught to navigate in the normative space of reasons. A psychology that rightly places its phenomena in the space of reasons begins from the fact that humans are able to normatively connect reasons as they appear in situations with appropriate actions. Adult human beings are able to give reasons for what they do, and they expect to be given such reasons from others in return. They are often able to evaluate if their actions (and thoughts and feelings) are based on *good* reasons, and they are sometimes able to change their desires if deemed undeserving in the light of worthier reasons. Others have talked about the space of reasons in which mental life is lived as "practical reality" (Dancy, 2000), a "moral space" (Taylor, 1989), "the local moral order" (Harré & van Langenhove, 1999) and "the moral ecology" (Brinkmann, 2004b). However, the basic idea is the same: If we do not think about human life as lived in some normative space, then we may be able to do physiological studies or maybe even sociological studies of human behavior, but we will never grasp the essential features of human psychology.

By locating mental life in the space of reasons, we may be able to depsychologize psychology. In the 1960s, Stanley Cavell wrote the following about Wittgenstein's famous book, *The Philosophical Investigations* (Wittgenstein, 1953), which also plays a considerable role in the present study:

> We know of the efforts of such philosophers as Frege and Husserl to undo the 'psychologizing' of logic (like Kant's undoing Hume psychologizing of knowledge): now, the shortest way I might describe such a book as the *Philosophical Investigations* is to say that it attempts to undo the psychologizing of psychology, to show the necessity controlling our application of psychological and behavioural categories; even, one could say, show the necessities in human action and passion themselves (Cavell, 1969:91).

This book is engaged in the project of depsychologizing psychology by arguing that psychology's phenomena are essentially normative and moral, located in the space of reasons.[6] And in the second part of the book, I argue the further point that the normative difference between *good* and *bad* reasons (for action, feeling, and thought) cannot reasonably be construed as subjective, as a matter of the individual's personal preferences. I defend the realist view that our moral reasons are given to us by the state of things around us and the value of these things, not by subjective beliefs, likings, or desires.

I thus try to develop a depsychologized psychology, i.e., a psychology that accepts that psychological phenomena are inherently normative. This kind of psychology will not feel threatened – but rather encouraged – by Alasdair MacIntyre's provocative claim that "Psychologies […] express and presuppose moralities" (MacIntyre, 1988:77). It will see the interpenetration of psychology and morality as a promising starting point for investigations into how all psychological processes – our patterns of feeling, desire, satisfaction, thinking, perceiving, and acting – are given meaning by existing in some particular "set of norms of justifications" (p. 76). Understanding psychological phenomena necessarily demands awareness of the "evaluative background," as MacIntyre says (p. 77) that constitutes the psychological phenomena as such. For example, we cannot make sense of the reaction (sadness) described in the episode above, if we do not know something about the evaluative background of the culture in which people watch sad movies and are moved emotionally by them.

[6] My claim that psychological phenomena are rightly placed in the space of reasons should not be taken to imply that we always have well-articulated reasons for what we do. In most situations we do not have conscious aims and corresponding justifications. I want to avoid the intellectualist fallacy that Bourdieu has underlined: "Very often researchers, because they are inspired by a will to demystify, tend to act as if agents always had as an end, in the sense of goal, the end, in the sense of conclusion, of their trajectory" (Bourdieu, 1998:82). We should not think that agents always "have reasons to act and that reasons are what direct, guide, or orient their actions" for "Agents may engage in reasonable forms of behaviour without being rational" (p. 76). This I fully accept, but still, I think it is undeniable (and I don't think Bourdieu would have denied this either) that social and psychological life depends on our discursive practices of giving and receiving reasons for action, and that without such practices we could not have what we consider full-fleshed human mental life. When Aristotle defined the human animal as a *rational* animal (a *zoon logon echon*) he did not mean that we always act rationally, but that what defines us is our capacity for conceptual thinking and speech. Heidegger (1927:47) followed Aristotle on this point and noted that *zoon logon echon* should be taken to mean "that living thing whose Being is essentially determined by the potentiality for discourse," but this does not lead to the (faulty) view that we always have reasons for what we do.

Aristotle as Background Figure

It is important to bear in mind that much of this normative approach to psychology is "old news" in Western thought and was thoroughly examined and articulated by Aristotle. In his study of Aristotle's psychology, Daniel Robinson says that "Aristotle's 'human science' is a *characterology*, a theory of 'personality' as today's psychologists would call it" (Robinson, 1989:94). Aristotle's developed psychology is not found in his *On the Soul*, but in his practical works, notably the *Ethics*. *On the Soul* is not a psychological treatise in the contemporary sense of the term, but a biological treatise about the living. It is a *psyche-logy*, but the *psyche* is not a "mind" in the modern sense, but the first principle of living things (p. 45). In the *Ethics*, however, Aristotle is concerned with the human being as an intentional creature whose operations demands teleological explanation. In the terminology used in the present book, he is concerned with the human being as a minded creature who lives in a normative space of reasons, where human action cannot be grasped in a causal framework.

Although Aristotle understood motivation (to take a classic psychological subject) as a natural phenomenon, he did not think that it could be fully understood by natural scientists (the *phusikos*). We also need the work of the "dialectician" in order to grasp motivation (Robinson, 1989:81). For the latter "would define e.g. anger as the appetite for returning pain for pain, or something like that, while the former would define it as a boiling of the blood." (Aristotle quoted from Robinson, 1989:81). The dialecticians place anger in the space of reasons, and know that there is such a thing as *justified* anger in the face of preposterousness. What makes "boiling of the blood" (or some modern neurophysiological equivalent) *anger* is precisely that it is situated in a practical context where it makes sense to question, justify and state the reason for "boiling of the blood." Anger is thus a psychological phenomenon in so far as it is a moral phenomenon, subject to praise and blame. If it were entirely outside the space of reasons, we should confine it instead to the science of physiology. As Harré (1983:136) has noted, the reason why dread and anger are psychological phenomena (i.e., emotions) but not indigestion or exhaustion – although all have behavioral manifestations as well as fairly distinctive experiential qualities – is that only the former fall, *for us*, within a moral order.

Harré says "for us," since he believes that classifications of what does or does not belong in the moral order (the space of reasons) are culturally relative, which means that what counts as a psychological phenomenon is culturally relative. I agree, but with the significant caveat (which Harré also endorses elsewhere) that some core features of human interaction that are psychologically basic seem to defy cultural relativization (later in the book I follow Anthony Holiday (1988) and analyze these as "core language games"). In the interpretive-pragmatic framework, there are non-constructed conditions for social constructions or, as Robinson explains Aristotle's human science: "Aristotle put forth a species of social constructionism, but one limited by realistic ethological considerations and the unique problems created by a self-conscious creature able to give and expect reasons for actions" (Robinson, 1992:97). That "man is taught by the *polis*" (*polis andra didaska*), is a premise in Aristotelian "social constructionism," but there are *objective moral values* that must be in place for the *polis* to teach humans anything. Thus a developed human psychological life presupposes a non-arbitrary normative moral order. In short, to repeat my two central theses: Psychological phenomena are normative, but not all normativity is conventional (and morality is one important kind of non-conventional normativity).

On the Notion and History of "Moral Science"

I employ the term "moral science" to direct us to the inevitable connections between psychology and morality. Unfortunately, the large majority of psychologists are either not aware of this connection, or simply ignore it in the name of what they think is value-neutral Science. But psychologists have not always been so wary of normative issues. In the nineteenth century, psychology belonged to what was then known as "the moral sciences," and in James Mark Baldwin's classical *Dictionary of Philosophy and Psychology* (Baldwin, 1901), the moral sciences are defined as:

> Those branches of inquiry which deal with mind and conduct, as opposed to matter and life; i.e. they are contrasted with the physical and natural sciences [...], and are often described as the 'mental and moral sciences'. In this general division all knowledge of man, apart from his body and its history, falls to the moral sciences; history, political economy, law, and statistics, as well as psychology, anthropology, and ethics.

It was the empiricist John Stuart Mill who first coined the term "moral science" in 1843. Mill's '*The Logic of the Moral Sciences*' (Mill, 1843), which was originally part of his *magnum opus*, *A System of Logic*, but subsequently published separately, was translated into German in 1849 by Schiel, and the word he chose was *Geisteswissenschaften* (Kessen & Cahan, 1986:649).[7] In this sense, the history of the moral sciences is short, almost as short as the history of psychology as an independent discipline. But in a broader sense of "moral sciences" as *practical* sciences the history goes back to Aristotle, and it is this broader sense that I wish to retain in the present study.

Before the advent of the modern scientific worldview and its theoretical underpinnings in the physical sciences, the practical sciences dominated European universities. The practical sciences – often conceptualized as "moral philosophy" – were *ethics*, the study of the nature of the good man, *economics*, the study of the good head of the household, and *politics*, the study of the good citizen, magistrate or prince (Smith, 1997:67–68). The view of the moral sciences in the medieval and renaissance universities was, in a sense, in direct continuation of Aristotle's classical project in his

[7] *Geisteswissenschaften* is thus not an old word, as many think, but the mid-nineteenth century German translation of Mill's "moral sciences." Of course, Hegel had conceived of his *Phenomenology* as a Science of Spirit – or, in German: a *Geisteswissenschaft* (Hegel, 1977) – but this is not yet the plural form *Geisteswissenchaften*, as used in the sense of the social sciences, e.g., by Dilthey (1977) in the later hermeneutic tradition.

practical works – the *Ethics*, *Politics*, and *Rhetoric*. Here, social inquiry and moral reflection were inseparable, and we can use the term "moral sciences" interchangeably with "social sciences" when we ponder Aristotle's practical works (Bellah, 1983). In these works, Aristotle formulated a version of the phenomenological view that "theory" is not the crux of the social sciences. According to Aristotle, the task of the social or practical sciences was not primarily to give us a *theory* about the good life (in the case of ethics) or a *theory* about the just *polis* (in the case of politics). Rather, it was the practical task of *making people good* and enabling citizens to live and reason excellently together in their communities. This conception survived in the medieval times and the renaissance, for example in the humanism of Erasmus and Montaigne (Toulmin, 1990, 2001).

But in the seventeenth century, at the beginning of the modern era, the pendulum swung, and the practical sciences were now viewed as inferior compared with the new and powerful theoretical abstractions of Science (Toulmin, 2001:29). As Stephen Toulmin recounts, theoretical *rationality* then superseded practical *reasonableness* in academic discussions, and moral philosophy went from being an intrinsic part of the disciplines, addressing all aspects of human practical life, to become a separate theory, discussed in abstraction from real life contexts.[8] For sixteenth century scholars, the paradigm of human reason was still a practical one as instantiated in the fields of law and jurisprudence, but for those working in the post-Galileian era, it became Science and Theory (Toulmin, 1990:34). The practical sciences, which, according to Toulmin's analysis, had been concerned with the particular, the local, the timely, the oral, and, I might add, the normative, were replaced by theoretical Sciences that favored the universal, the global, the eternal and the written word.[9]

Concerning psychology more specifically, it is the case that this discipline originally emerged from a particular moral discourse that flourished in eighteenth century Britain. According to Kurt Danziger, this moral discourse was "based on a fundamental sense of separation between human individuals as well as between individual agents and their actions" (Danziger, 1997a:181). Yet when modern psychology wanted to become a respectable science, it felt that it had to separate itself from moral issues (Graumann, 1996). This separation can be seen in a large number of conceptual changes. A term like "behavior" originally belonged in a discourse of moral praise and blame (Danziger, 2003) (like when we say "behave yourself!"), but later became a morally neutral concept

[8] Toulmin (2001:135) confers the rise of this kind of "moral theory," separated from other kinds of inquiry, to Henry More and the seventeenth century Cambridge Platonists. The Aristotelian and medieval focus on casuistry, rhetoric, and practical reasonableness was then replaced by a focus on abstract theories.

[9] Even though the practical sciences were superseded in the course of the seventeenth and eighteenth centuries by theoretical sciences, subsequent scholars in such fields as economics still prided themselves with belonging to "moral philosophy." Adam Smith, moralist of the Scottish Enlightenment, and political economists like Jeremy Bentham and later John Stuart Mill "systematically discussed motivation and ethics as well as economic realities. Their science of man was a moral philosophy," as Roger Smith has put it (Smith, 1997:317).

with behaviorism and contemporary forms of behavior analysis. Early methods of personality re-formation called "moral therapy" were transformed into the value neutral notion of "psychotherapy" (Charland, 2004). A journal like *Character and Personality* became *Journal of Personality* as late as 1945 in order to avoid the moral connotations of "character" (Greer, 2003:97). "Moral insanity" became "sociopathy" or "personality disorder." The conceptual and discursive shifts that occurred around the eighteenth and nineteenth centuries were so massive that psycho-historian Danziger (1997a) talks about this epoch as "the Great Transformation" that paved the way for modern, non-moral psychology.

The "Everywhereness" of Psychology

Mark Jarzombek, an architect who writes about the impact of psychology on modern art, talks about the "everywhereness" of psychology in Western culture (Jarzombek, 2000:12). He argues that psychology is for the modern age what "perspective" was for the Renaissance: something, which, after its emergence, has left no aspect of our world untouched. The West has been thoroughly psychologized, he argues, and psychology's main field of operation has not been "in the controlled environments of the scientific laboratories of Wundt and Lipps, but in the more free-wheeling discursive practices of philosophers, historians, avant-garde artists, cultural reformers, and politicians" (p. 16).

Roger Smith, author of the voluminous *Norton History of the Human Sciences* correspondingly finds that contemporary Western societies have become Psychological Societies (Smith, 1997), and this will play a key role in the present study. Psychological discourses have come to play an important role in our lives, and have made people understand themselves according to the vocabularies, theories, practices, and techniques of modern psychology. Examples are easy to think of. Psychotherapeutic practices have today become a commonplace. Not just in the form of ordinary therapy where one goes to see the therapist in order to solve a problem, but also in talk-shows, radio phone-ins, magazines, self-help literature, employee interviews, educational counseling, coaching, and also private conversations. Besides therapeutic practices, other kinds of psychological practices have flourished in the West. Mental tests have become a *sine qua non* in job interviews, school curricula and assessments of criminals. Psychologically informed self- and personality-development courses have entered not only our private lives, but also the domains of work and education (Illouz, 2007). Comprehensive ethnographies could be written to portray the psychological people of the West, who tend to think of themselves as psychological subjects with an inner, psychological realm, and who employ specialized psychological techniques to give shape to their selves (such work is emerging today, e.g., in Jansz & van Drunen, 2004; Rose, 1996a, 1999a; Smith, 1997).

In the analyses that follow, I conceptualize the "everywhereness" of psychology in the West by expanding on Charles Taylor's concept of "the social imaginary" (Taylor, 2004). I argue that the lives of most of us embody a *psychological social imaginary* (see especially Chap. 2). We tend to imagine social life and the moral order in psychological terms, which, I argue, has made it difficult for us to act and orient ourselves adequately in moral space. We tend to think about the moral values and reasons that we live by as interior, psychological properties rather than as aspects of the social world in which we live. Often, these interior properties are explained with reference to the evolution of the human species, for example in the currently influential field of evolutionary psychology. Morality is often hereby *psychologized*, transformed from external normative demands into a pure psychological phenomenon, which means that the normative dimension of morality is routinely defined away, and only the descriptive dimension is left behind. I shall argue that such psychologization is wholly inadequate and makes moral phenomena unrecognizable as such. Rather than psychologize morality, this book suggests that we ought to "moralize" psychology – or at least realize that psychology is always already "moralized!"

Psychological Representations of Morality

At this point, the reader may ask if it is really true that morality is routinely neglected as a fundamental human phenomenon in psychology. I grant that although most psychologists do not study morality, some do, but my point is that they normally proceed in ways that are inadequate to grasp the normativity of moral (and psychological) phenomena. From a bird's eye perspective, there are three standard ways in which morality is addressed in psychology. Morality is either (1) ignored, (2) reduced (to biology or social constructions), or (3) narrowed into rules.

1. Morality as virtually non-existing
   In several major handbooks of psychology, morality is more or less treated as non-existing. *The International Handbook of Psychology* (Pawlik & Rosenzweig, 2000), which is supposed to cover "all the main areas of psychological science" (as promised on the book cover), treats "morality" in just two pages out of more than 600, viz. under the heading of "moral development," where Kohlberg serves as the major reference. The case is similar with *The Corsini Encyclopedia of Psychology and Behavioral Science* (Craighead & Nemeroff, 2001) and *The Gale Encyclopedia of Psychology*, whose only references to moral issues likewise concerns "moral development," which is defined in a Kohlbergian fashion as "The formation of a system of underlying assumptions about standards and principles that govern moral decisions" (Strickland, 2001:436). That morality is given a very narrow treatment as a developmental phenomenon, *ad modum* Kohlberg, is quite typical in psychology, which makes it invisible as a pervasive phenomenon in our everyday lives.

2. Morality as reducible
   Kazdin's *Encyclopedia of Psychology* also treats morality as a developmental issue only, but it does include an extensive discussion of the concept of "values."

Values are defined as "beliefs pertaining to desirable end states or modes of conduct that transcend specific situations, are organized into coherent systems, and guide selection and evaluation of people, behaviors, and events" (Kazdin, 2000, vol. 8:153). The definition reduces the normativity of values into descriptive *beliefs*. Normally we would reject the view that values are beliefs, because that would mean that values would change when our beliefs change. In our everyday lives it is rather the case that we evaluate our beliefs in the light of values and moral reasons (e.g., when we ask: "do I have a good reason to believe so and so?"). Thus, it is problematic to say, like Kazdin's reference, that moral values can be understood purely descriptively as beliefs.

Such reductions of morality are often quite implicit and non-reflected in the psychological literature, but, at other times, psychologists self-consciously believe that morality can and should be reduced to something else. In this book, this group is primarily represented by a number of evolutionary psychologists and sociobiologists, whose arguments are discussed and ultimately rejected in Chap. 6. Some of these consider morality an epiphenomenon of our biology, something whose ultimate purpose is to keep the genetic material intact. We may believe that morality is objective, but, their argument goes, this belief is itself the result of natural selection; it has proven to be adaptive and therefore chosen for (Ruse, 1991). This subjectivist view is quite subtle and a significant challenge to the interpretive-pragmatic form of moral realism that is defended in this book. Another reductionist moral theory in psychology can be found in B.F. Skinner's behaviorism. Skinner argued that the predicate "is good" should be defined as "is reinforcing." Good things are positive reinforcers, he claimed (1971:96), which means that the moral imperative to love your fellow human beings can be reduced to two descriptive facts: "(1) 'The approval of your fellow men is positively reinforcing to you' and (2) 'loving your fellow men is approved by the group of which you are a member'" (Skinner, 1965:429).

A very different kind of reductionism that will also be discussed in Chap. 6 (as well as in other chapters) is social constructionism. A common variety of this view reduces morality, not to biological survival, but to the prevailing and contingent social opinions (or constructions). Morality is thus presented as something wholly relative, and, in Kenneth Gergen's version, as a pure linguistic construction. Gergen argues that moral languages are "moves or positionings that enable persons to construct the culture in what we take to be a moral or ethical way" (Gergen, 1992:17). There is nothing beyond these languages for them to be true to, and, in that sense, the possibility of us being in error, morally speaking, cannot arise. This is a version of moral anti-realism that portrays morality as a positioning device, which serves individual needs. I argue against this view by pointing to what I believe are certain universal moral conditions that must be in place for the whole social constructionist account to make sense. These conditions are real and objective, and cannot be accounted for as contingent or socially constructed.

3. Morality as rules
   The most common way of theorizing morality in psychology is to understand it as a system of rules. This was given a clear articulation by Jean Piaget in his classic work, *The Moral Judgment of the Child* (Piaget, 1932), which was

inspired by Kant's moral philosophy. "All morality consists in a system of rules, and the essence of all morality is to be sought for, in the respect which the individual acquires for these rules," says Piaget (p. 1). Piaget's approach was further developed by other "Kantian" psychologists such as Lawrence Kohlberg and Elliot Turiel (Kohlberg, 1984; Turiel, 1983), whose works I address in Chap. 7, and it is still dominant in today's reference books and encyclopedias, as we saw above. It is telling that Darley and Schultz's comprehensive review of "modern moral psychology" had an exclusive focus on "morality as rules," which they believed was in line with the prevailing information processing paradigm in cognitive psychology (Darley & Schultz, 1990). The current view of the mind as an algorithmic processor of information suits a rule-based morality very well, and vice versa. The idea is that, ultimately, morality can be codified, and the task of moral psychologists is to investigate how humans internalize and learn to follow the given set of moral rules (Lapsley, 1996:16). The interpretive-pragmatic view developed in this book does not as such reject the relevance of rules in moral life, but it argues that moral rules are to be thought of as tools, which are not authoritative in themselves, but valid only to the extent that they help us act adequately in concrete situations and practices.

Against these standard treatments, or non-treatments, of morality in psychology, the interpretive-pragmatic view developed in this book presents the following arguments:

1. Morality is something objectively real that co-constitutes psychological phenomena and therefore should not be ignored by psychologists. In fact, if there is a bedrock to psychology, it will be found in an elementary moral normativity that orders our lives and not in any other psychological phenomena, most of which are much more susceptible to cultural and historical relativity.
2. Morality is irreducible since it cannot be accounted for in pure non-moral terms. If psychological phenomena are closely aligned with moral ones, then this implies that psychological phenomena themselves are irreducible (e.g., to neuroscientific explanation).
3. Morality cannot be codified into a set of rules, for moral phenomena are always embedded in situations that call for interpretation, imagination, and judgment; social practices rather than rigid rules are therefore the chief source of moral intelligibility.

The Structure of the Book

I believe we have at least two good reasons to include psychology among the moral sciences again (although, as this book will show, we ought to think differently about what "moral science" is than most nineteenth century writers). These two reasons structure the two parts of this book:

1. Psychology today is an important cultural agent in "making up people," and therefore has *moral effects* on our self-understandings and practices (Part I: Chaps. 2–4).

2. Psychology's subject matter is irreducibly moral; psychology has *moral contents* (Part II: Chaps. 5–8).

The first reason stems from the "everywhereness" of psychology in contemporary Western societies. Psychology continues to influence how we think about the "oughtness" of social existence by fabricating normative standards about how to be, act, think, feel, and live, and it provides people with tools (e.g., therapies, pedagogies, and tests) to reach these standards. Iris Murdoch has said: "Man is a creature who makes pictures of himself and then comes to resemble the picture. This is the process which moral philosophy must attempt to describe and analyse" (Murdoch, 1997a:75). Psychology is an important, and perhaps the *most* important, picture-maker of humans today, and, as Murdoch points out, it is indeed a *moral* task to analyze the process in which pictures are fabricated and humans are made in their image. I follow Ian Hacking (1986; 1995b) and refer to this process as "the looping effect of human kinds" (a central theme in Chap. 4). If psychology has the power to fabricate human kinds that make up people (Hacking, 1986), then psychologists have a good reason to examine the ethico-political presuppositions and implications of their discipline. Psychology is both embedded *in*, affected *by*, and in turn *affects* our moral space and our social imaginary.

The other important reason we have to conceptualize psychology as a moral science stems from the claim (to be defended) that humans are irreducibly moral beings. An adequate science of such beings must correspondingly be a moral one. The claim springs from the argument that mental life is lived in the space of reasons. Humans are intentional creatures whose lives incorporate attempts to realize certain goals, and we are to some extent able to evaluate whether our goals are *worth* striving for, and we can sometimes give *reasons* for our goals, thereby *justifying* our actions. Rather than accommodating our morality to what we think we know about human psychology, I believe we should begin with undeniable moral features of human experience and develop a psychology that respects our moral abilities (i.e., our abilities to perceive moral features of the world, to evaluate our reasons for action, and to strive to become moral persons of a certain standing). As we will see, these two reasons are themselves related, for the looping-effects are possible *because* of the fact that humans are intentional and self-interpreting creatures, and it is also this that makes our existence as moral beings intelligible.

Part I of this book examines the moral consequences of the psychologization of the West, looking for "the place of value in a world of psychology" (cf. Köhler, 1959). Part II turns around and argues that morality is an inescapable part of what psychologists study.

Part I
The Place of Value in a World of Psychology

Chapter 2
The Psychological Social Imaginary

In the introductory Chapter, I argued that psychological modes of understanding are pervasive in today's Western culture. Roger Smith concludes that modern society is a Psychological Society. In the twentieth century, "everyone learned to be a psychologist, everyone became her or his own psychologist, able and willing to describe life in psychological terms" (Smith, 1997:577). In this chapter we turn our attention to the roots of Psychological Society with a special focus on the effects of psychologization on our moral lives. How have we learned to think about morality in an age dominated by psychological modes of understanding? How is it even possible to think about morality from the perspective of a psychological worldview? In order to answer these questions, we need to know what "the psychological worldview" is, how it arose historically, and how this worldview relates to our conceptions of morality and normativity.

The twentieth century was not only a psychological age, but, according to leading moral philosophers (MacIntyre, 1985a; Taylor, 1989), also an age when morality became subjectivized.[1] Moral subjectivism is the view that something is morally good if and only if the moral agent has a positive preference towards it. In philosophy this is an old view, but propounded with particular force by David Hume in the eighteenth century. In the twentieth century, moral subjectivism was heralded in academia under the name of emotivism, which, according to Alasdair MacIntyre (1985a), became embodied in the Western culture, its sciences, institutions and discourses. In the twentieth century, the West simultaneously witnessed a psychologization of society and a subjectivization of morality. This chapter investigates the relations between these two processes.

[1] I distinguish between "subjectivization" as the process where something becomes internalized, finding its source in the "inner world" rather than the outer, and "subjectification" as the process where humans are made subjects in specific ways. The latter term owes much to Michel Foucault and his analyses of "the different modes by which, in our culture, human beings are made subjects" (Foucault, 1994b:326), which will be discussed in later chapters. My claim is that psychology has been equally involved in the subjectivization of morality and the subjectification of human beings.

S. Brinkmann, *Psychology as a Moral Science: Perspectives on Normativity*,
DOI 10.1007/978-1-4419-7067-1_2,

I approach the modern processes of psychologization and subjectivization in terms of what has been called the "social imaginary." In his book on *Modern Social Imaginaries*, Charles Taylor defines social imaginary as

> the ways people imagine their social existence, how they fit together with others, how things go on between them and their fellows, the expectations that are normally met, and the deeper normative notions and images that underlie these expectations (Taylor, 2004:23).

Taylor wants to stress the fact that the social imaginary is not simply the cluster of intellectual ideas we employ when we think about social relations. It is not an explicit social theory, but rather what determines how we formulate such theories. It determines which questions we can meaningfully ask about our social existence (and which we cannot ask), and it affects the ideas we form. Taylor uses the term "imaginary" because his focus "is on the way ordinary people 'imagine' their social surroundings, and this is often not expressed in theoretical terms, but carried in images, stories, and legends." (Taylor, 2004:23). Taylor draws in three social forms that play a significant part in the modern social imaginary: the market economy, the public sphere, and the self-governing people (p. 2). In addition to these, I shall argue that psychology, as an array of practical modes of understanding and acting, should be seen as having penetrated our social imaginary to the extent that we have problems seeing that social life can be imagined in non-psychological terms, and indeed was historically imagined in other terms prior to the eighteenth, nineteenth and twentieth centuries (and perhaps is still imagined so today in certain non-Western societies). When I address "psychology" in what follows, I do not merely refer to the academic discipline, but to a whole cultural form; a culturally specific way of understanding and ordering actions. This was also Foucault's understanding of psychology:

> I don't think we should try to define psychology as a science but perhaps as a cultural form. It fits into a whole series of phenomena with which Western culture has been familiar for a long time, and in which there emerged such things as confession, casuistry, dialogues, discourses and argumentations that could be articulated in certain milieus of the Middle Ages, love courtships or whatnot in the mannered circles of the seventeenth century. (Foucault, 1998b:249).

In later chapters I present a view of social practices as the background that enables things and situations to appear as meaningful. The social imaginary, however, should be understood as even more basic than specific practices. The social imaginary is that common understanding that makes possible shared practices in the first place (Taylor, 2004:23). Our practical background understandings, our implicit knowledge of what to do in different situations, would not be possible without "a wider grasp of our whole predicament: how we stand to each other, how we got to where we are, how to relate to other groups, and so on" (p. 25). The notion of the social imaginary is intended to capture this "wider grasp of our whole predicament" that grounds even our background understandings of practices. Social practices are not isolated islands, unconnected ways of doing things, for most practices only make sense in their relations to other practices, and there is often a common cultural form to how people comport themselves in different social practices. Practices are held together by a common understanding and this common understanding is what the notion of the social imaginary is supposed to capture.[2] In short, the social imaginary is our "implicit grasp of social space" (p. 26), and my argument is that this implicit grasp of social space has increasingly been psychologized since the eighteenth century.

In relation to morality, there are at least two interesting features of the concept of social imaginary. First, an important part of the social imaginary is "a sense of moral order" (Taylor, 2004:28). The social imaginary incorporates some sense of how we all fit together in shared social practices, and such understanding is simultaneously factual and normative (p. 24). The social imaginary concerns the *oughtness* of practical life. Second, although it is impossible to make the social imaginary fully explicit in theoretical propositions, it is nonetheless susceptible to being influenced by otherwise explicitly formulated theories and ideas (p. 28). This makes the concept useful for an investigation of how psychological ideas have affected our social and moral lives.

The Two Faces of Psychology

It is easy to claim that psychology has infiltrated our social imaginary, but it is hard to pinpoint what psychology really is. Although it is clearly not a simple and unitary thing, but an amalgam of different theories and practices, we may, however, discern a unity in the otherwise "fragmented and contradictory field of psychology" (Kvale, 2003:596). The diverse languages of psychology can be seen as united in its *Janus head*, talking with two tongues: One face of the Janus head presents exciting therapeutic narratives and vivid accounts of personal change and development, "legitimating a psychology of human concerns" (p. 596). The other face of the head speaks the language of statistics and quantitative experiments and questionnaires, "legitimating psychology as a natural science" (p. 596). The two faces rarely talk to or confront one another, for each is dependent on the other in spite of their incommensurable natures. One face gives the discipline its scientific legitimacy, funding and academic prestige and positions, while the other gives practical relevance and entertains the public in Psychological Society. What the two faces share is a common conception of the individual as the basic unit of psychology.

In this chapter, I will argue that both faces of psychology's Janus head have shaped our social imaginary. I try to trace the roots of the faces to two strands of eighteenth

[2] The concept of social imaginary resembles Foucault's (2001) notion of *episteme*, which notably figured in the early parts of his work. But while Foucault understood the *episteme* as something like an unconscious cultural code to be made explicit by structural analysis, Taylor rejects the idea that the social imaginary can be fully expressed in explicit doctrines. It is *lived* rather than *thought*, based on habitual, bodily practices rather than underlying social rules.

century thought: modernism and romanticism. Psychology may have been born in 1879 when Wilhelm Wundt founded his laboratory in Leipzig, but it was already conceived in the first half of the eighteenth century. Kurt Danziger (1997a) has described the eighteenth century as the Great Transformation, the century when psychology found its language.[3] According to Danziger, psychology then not only found a *language* suitable for representing a pre-existing realm of psychological phenomena. He argues more controversially that it also created its *phenomena*: "Before the eighteenth century there was no sense of a distinct and identifiable domain of natural phenomena that could be systematically known and characterized as 'psychological'" (p. 37). There were theological, philosophical, moral, medical and political phenomena, but no psychological phenomena. The relations of humans to their world, the deity, their bodies, and their fellow human beings were not yet imagined in psychological terms. This is not to deny that people reflected on their experiences before the advent of psychology, but rather to insist that such reflection took on a new meaning and was structured differently after the introduction of psychology.[4]

Instead of talking about modernism and romanticism in abstract and general terms, I turn to David Hume (1711–1776) and Jean-Jacques Rousseau (1712–1778) as two contemporaneous thinkers, who exemplify these strands of thought, and whose ideas have been formative in relation to the two faces of the Janus head of contemporary psychology.[5] From Hume's modernism came the idea that the world, and human behavior within it, could be understood in total as a mechanical system with the help of scientific methods. Hume thus wanted to "introduce the experimental method of reasoning into moral subjects," as the subtitle of his first book from 1739 proclaims (Hume, 1978). Hume's early psychology is an attempt to instigate a science of the mind on Newtonian premises, and Hume can thus be presented as the grandfather of modernist psychology, as a precursor to the scientific face of

[3] It was also the century when the term "psychology" gained a usage. According to Raymond Williams's *Keywords* (Williams, 1983), the word "psychology" entered the English language in the seventeenth century in the sense of "a doctrine of souls," but in the scientific sense of "empiric psychology," the word was first used by Hartley as late as 1748, where he took up Wolff's German definition from 1732. Williams adds that the word was not much used before the nineteenth century.

[4] In his archeology of the human sciences, Michel Foucault was even led to claim that: "Before the end of the eighteenth century, *man* did not exist" (Foucault, 2001:336). This does not mean that human beings did not exist, but that "man," as a privileged object of research, did not exist. Foucault also touches upon the advent of psychology, and argues that "the new norms imposed by industrial society upon individuals were certainly necessary before psychology [...] could constitute itself as a science" (p. 376). In *The Order of Things*, Foucault introduces another of the themes of this book in his declaration that "Modern thought has never, in fact, been able to propose a morality" (p. 357). The reason for the inability of modern thought (psychology included) to propose a morality is that for modern thinkers, "any imperative is lodged within thought" (p. 357), i.e., within the thinking subject rather than in "the order of the world."

[5] Hume and Rousseau were in fact personal friends, and when Rousseau had to flee from France – where his book *Émile* was burned in public immediately after its publication in 1762 – Hume arranged for him to come to England.

psychology's Janus head. From Rousseau's romanticism came the idea that humans possess a deep interior to be unleashed through a process of self-realization.[6] This idea has been important to twentieth century humanistic psychology, and more generally to the public face of psychology's Janus head. Rousseau provided later psychologists with a subjectivist language of human concerns that enabled psychology to become a secular technology of self-realization.

Hume and Rousseau, in spite of their many differences, agreed on some fundamental points: Notably concerning (1) a shared focus on the world of private experiences, (2) a shared moral subjectivism, and (3) a shared atomistic view of society as something established through individuals' contractual consent. These three elements incarnate what I will call the psychological social imaginary. As we shall see, Hume's focus on *method* as the key to a science of the mind, became formative in psychology, and the primacy of method penetrated into the social imaginary and human self-understanding through a process that I shall call "ontologizing" (see also Taylor, 1993): "Method" was read into the very constitution of the mind itself, thereby contributing to shaping humans in light of a method-based scientific psychology (more on this below). Rousseau's romantic ideas of the inner self and its realization have run in tandem with the modernist focus on method, and, together, they have enabled psychology to become an active participant in turning humans into specific kinds of psychological subjects in a value-free and disenchanted world.

The Worldly Nature of Psychologization

Initially, however, I should subject the approach of the present chapter to self-criticism. There are at least two complementary limitations of my approach that are worth mentioning: First, why begin an investigation of the history of psychologization in the eighteenth century and why choose Hume and Rousseau (none of whom were psychologists in a modern sense of the term)? Second, isn't the approach blatantly idealistic? Why recount the history of psychologization in terms of thinkers and their thoughts? Why not in terms of the development of concrete historical practices, where psychology has been connected with social management and a political interest in controlling individuals and populations? Should I not instead do a social history of psychology (Jansz, 2004) or what Foucault (1998a), referring to Nietzsche, called *wirkliche Historie*?

As regards the first point, I concede that it is unorthodox to argue that much of what is interesting about psychology's history and its current Janus-head situation can be traced to ideas from the eighteenth century, and to such authors as Hume and Rousseau. Conventional histories of psychology typically begin with Descartes, or

[6] I do not think Rousseau himself used this word, but there is agreement among interpreters that his philosophy amounts to (and indeed inaugurated) a form of self-realization thinking (Wokler, 2001).

even with the ancient Greeks. But such histories ignore the fact that, as Danziger says, "the very notion of 'psychology' in the modern sense, forming a distinct field of study, can hardly be said to have existed before the eighteenth century" (Danziger, 1997a:21). They uncritically assume that Aristotle's *psyche* corresponds to the Latin *anima*, to the Christian *soul* of the middle Ages, and to the *mind* of modernity. In my view, this cannot be taken for granted, since a number of historical, philosophical and ethnographic accounts have demonstrated that the modern idea of "the psychological," as an inner realm of thoughts and feelings, is "a function of a historically limited mode of self-interpretation, one which has become dominant in the modern West and which may indeed spread thence to other parts of the globe, but which had a beginning in time and space and may have an end" (Taylor, 1989:111). From an anthropological viewpoint, Clifford Geertz likewise observed that:

> The Western conception of the person as a bounded, unique, more or less integrated motivational and cognitive universe, a dynamic center of awareness, emotion, judgment, and action organized into a distinctive whole and set contrastively both against other such wholes and against its social and natural background, is, however incorrigible it may seem to us, a rather peculiar idea within the context of the world's cultures (Geertz, 1983:59).

Around the time of Hume and Rousseau, we find perhaps for the first time a way of talking about psychological phenomena that seems compatible with how we imagine the psychological domain today. Also the very term "psychology" was then introduced into English, and about 100 years later, the philosophical ideas of Hume and Rousseau were supplemented with the experimental practices of German physiologists, thus assembling the modern scientific discipline of psychology. Especially, the ideas of Hume made it intelligible to investigate the mind experimentally and methodologically in the manner of Wundt and Ebbinghaus. The mind had to be "imagined" as something susceptible to methodological investigation in order for experimental psychology to make sense, and it was Hume who most clearly articulated the necessary kind of imagination. Rousseau's focus on the inner voice and the development of the self likewise made it reasonable to imagine therapy, counseling, pedagogy, and business consultancy in the manner of Carl Rogers, for example, and most of today's psychological practitioners who put premium importance on the subjective "inner voices" of clients (Illouz, 2008).

I admit that this way of putting things could strike one as idealistic, which takes us to the second line of criticism. Critical psycho-historians argue that the discourse of psychology became necessary because of changed social and economical circumstances, particularly in eighteenth century industrialized Britain (Danziger, 1997a:181). "The psychological," argued cultural theorist Raymond Williams, emerged as a "great modern ideological system" that, with the beginning of industrial capitalism, began to make available new forms for structuring subjectivity (Williams, 1978:128–129). Roger Smith has also pointed to the practical and worldly nature of psychologization:

> subjects like psychology and sociology did not originate in the academic setting as much as in the administrative and institutional means developed to manage human beings [...] it was the schools, prisons, asylums, hospitals, workhouses, families, government reports, charities, church groups, youth movements, friendly societies and factories – the local day-to-day management of human activity – which turned man into a systematic object of study (Smith, 1997:374).

Furthermore, a number of commentators have observed that the social sciences – psychology among them – and the liberal nation state, including its mass national school system, emerged and overlapped historically (Christians, 2000:134). There is a historical co-emergence of the institutions of the modern nation states and the knowledge about individuals and populations produced by the social sciences, notably psychology. This in itself should alert us to the idea that the psychological way of thinking about humans cannot be morally and politically neutral. As Foucault argued, the historical emergence of social science cannot "be isolated from the rise of this new political rationality and from this new political technology" (Foucault, 1988b:162). Social science was needed in the new nation states to administer and govern. In Foucault's perspective: "the emergence of the human/social sciences is contemporaneous with, and indivisible from, the development of disciplinary power" (Hook, 2003:609).

I agree that it was in these practical contexts that the psychologization of the world occurred. But I also think that there is more to the story than "governmentality" and "disciplinary power" (I expand on the Foucauldian perspective in Chap. 4, where I also subject it to some criticism). I am particularly inspired by John Dewey, who was interested in the conditions that must exist in order for psychology to emerge and make sense. Dewey found that "if any individual is taken as a member of a limited social group, we cannot have and historically did not have any psychology as psychology. [...] as a science could not come to birth because the individual as a possible universal had not come to existence" (Dewey, 1976:4). Not until a certain freedom is granted in societies – with new constitutional state formations – does the individual come into existence and "becomes the object of a science – psychology" (p. 4).[7]

So although my story of the psychologization of the world largely works on the level of the development of *ideas*, these ideas should be thought of as embedded in *practices* and inevitably connected to cultural and societal realities. Here I use the word "practice" inspired by Taylor (1989), to denote "something extremely vague and general": "any stable configuration of shared activity, whose shape is defined by a certain pattern of dos and don'ts, can be a practice" (p. 204). The basic assumption is that "ideas articulate practices as patterns of dos and don'ts. That is, the ideas frequently arise from attempts to formulate and bring to some conscious expressions the underlying rationale of the patterns" (p. 204). Nikolas Rose says that ideas "are bound into ways of seeing and acting: into technologies. They are enmeshed in definite practices of experimentation, investigation, and interrogation arising not only in the laboratory or the academic's study but in an array of social locales" (Rose, 1996a:83). I share this perspective that regards ideas and theories as practical tools or technologies that operate in the world, and which do not stand apart from the world and passively represent it. As Dewey put it: "The so-called separation of theory and practice means in fact the separation of two kinds of

[7] Dewey's account owed much to Hegel's argument (1821) that individual subjects do not emerge in the course of history before complex social formations governed by a legal system come into existence (see Brinkmann, 2004a).

practice" (Dewey, 1922:69). Inspired by Taylor, I will insist on the need for rethinking the relations between ideas and material factors:

> what we see in human history is ranges of human practices that are both at once, that is, material practices carried out by human beings in space and time, and very often coercively maintained, and at the same time, self-conceptions, modes of understanding (Taylor, 2004:31).

When we trace how psychology has infiltrated our social imaginary, we should therefore equally notice how psychological ideas have formed our self-understandings on the one hand, and also how new psychological ideas have been responses to changing material practices on the other. We should abstain from deciding the direction of the causal arrow beforehand: "The only general rule in history is that there is no general rule identifying one order of motivation as always the driving force. Ideas always come into history wrapped up in certain practices, even if these are only discursive practices" (Taylor, 2004:33).

In what follows I give an outline of the "psychologies" of Hume and Rousseau (who were, of course, parts of broader intellectual traditions). I am aware that both thinkers were much more sophisticated than the following brief remarks might indicate. My analyses should be seen as sketchy ideal types; ways of thinking about humans that emerged with the transformation of the feudal order and the Church's authority, the disenchantment of the world, the rise of natural science, and the emergence of new nation states in Europe. All these worldly events form the background to the psychologization of the world that culminated in the twentieth century with the Psychological Society (Smith, 1997). In this context it is impossible to give a full historical reconstruction of psychology and psychologization. My presentation of modernism and romanticism – Hume and Rousseau – is intended as a second best option; as selective steps backwards in history to rediscover certain sources, which will hopefully illuminate how the two faces of psychology have shaped and in some ways distorted our conceptions of morality.

Hume: The Newton of the Mind

David Hume's philosophy was the culmination of a movement in Western thought that had been on its way for centuries, referred to, by Kessen and Cahan (1986:640), as "The great Western transcendental slide from God to Nature to Mind to Method." Hume's approach, and the later psychology influenced by it, was based, not on assumptions about God, Nature, or even Mind, but rather on Method, for Hume wanted "to introduce the experimental method of reasoning into moral subjects" (Hume, 1978:xi). In the case of Hume, the "basic science of human nature may properly be designated by the term 'psychology'" (Miller, 1971:155), although, as I remarked above, the term "psychology" was not introduced into English until 9 years after the publication of Hume's Treatise, namely in 1748 by Hume's contemporary, and fellow empiricist-associationist, David Hartley.

Hume was greatly impressed by Newton's mechanical physics, which had made it possible to comprehend the physical world in terms of universal laws of nature, rendering it calculable and, to some extent, controllable. It was in a mechanical, disenchanted world that Hume found himself, and it was here he set out to develop a Newtonian science of the mind. By introducing the experimental method of reasoning into moral subjects, Hume wanted to reform the science of man. He believed that all sciences, including mathematics and natural philosophy, depend on the science of man "since they lie under the cognizance of men, and are judged of by their powers and faculties" (Hume, 1978:xv). Hume would "in effect propose a complete system of the sciences, built on a foundation almost entirely new, and the only one upon which they can stand with any security" (p. xvi). Psychology is seen as the basic science, a first philosophy.

Like his empiricist predecessors, Hume carried forth Descartes' representational epistemology (the "spectator theory of knowledge" as Dewey later called it) according to which the mind, envisioned as a kind of container, is in direct contact with its given contents ("impressions" and "ideas" in Hume's words). Hume continued the Cartesian project of making epistemology the prime philosophical discipline, taking the lead from the question: How can I, as a discrete, thinking being, know anything about the external world? In many ways, this epistemological question has been psychology's main question as well, as Yanchar and Hill (2003) argue in their critique of the dominance of epistemology in psychology; a dominance that has largely excluded concerns with the very purpose and subject matter of the discipline. A discipline dominated by epistemological questions quite naturally concentrates on its methods rather than basic ontological questions, and this tendency, which Sigmund Koch once referred to as *epistemopathic* (Koch, 1981), can be traced to Hume.

For the purpose of the present investigation, the most important part of Hume's psychology is his understanding of morality. He clearly understood morality as a pure psychological phenomenon, grounded in nothing but natural sympathetic dispositions and reactions (Robinson, 2002:18). Hume is still unrivaled in his sophisticated version of moral subjectivism, and, according to Thomas Nagel, it is still the case that "The point of view to defeat, in a defense of the reality of practical and moral reason, is in essence the Humean one" (Nagel, 1997:106). In Hume's eyes, what we call morality is the result of a strengthening of relations between certain actions that are value-neutral in themselves and our subjective reactions in terms of pleasure and pain. To simplify: Those actions that I like are morally good, and those actions that I dislike are morally bad. There are no moral qualities and no values in the world. Moral qualities are subjective projections unto a value-neutral, mechanical world, and a description of the world in toto, given in value-neutral terms, is a complete description. Fortunately, Hume thought, there are common human tendencies to react to events such as murder in emotionally similar ways, which means that, as a matter of fact, our subjective projections of values unto situations tend to be similar. But this is a contingent psychological fact that could change. Hume's psychologization of morality implies treating moral values as psychological facts from a detached scientific viewpoint. And most of the later psychology, according to Leslie Smith's useful discussion, has in fact inherited what he calls Hume's "non-normative interpretation of norms," which, alas, completely bypasses what is essential to normativity, as we shall see (Smith, 2006).

The main Humean points that came to shape the later science of psychology can be summarized as follows:

Methodolatry

First, his call for a science of man to be based on the experimental method of reasoning echoed in much later psychology. The notion of methodology became important to unite the otherwise fragmented discipline of psychology, but the stress on method also served to exclude morality and values from most psychological inquiry: How could methods that were designed to investigate *facts* teach us anything about *values*? They could not, and have not, except when psychologists have treated values as unproblematic facts in accordance with Hume's subjectivist projection-theory of value. In Hume's psychology, only passions can motivate; reason can merely calculate the optimal means to reach the ends dictated by passion (Danziger, 1997a:44). Reason, as Hume famously said, "is, and ought only to be the slave of the passions, and can never pretend to any other office than to serve and obey them" (Hume, 1978:415). The modern instrumental and method-based view of rationality was clearly articulated in Hume's philosophy.

Subjectivism

Second, Hume referred all that is existentially important in people's lives – meanings, values, morals – to the subjective realm of the mind, albeit in Hume's version this was a realm without a sovereign self.[8] The outer world seemed to Hume to be explicable in Newtonian terms as a mechanical universe, and so he needed a corresponding mechanics of the mind – a psychology – to explain such things as meaning and morality. In this regard, Costall (2004a:184) has described psychology as a mistake waiting to happen: "When physical science has promoted its methodology (of atomism, mechanism, and quantification) to an exclusive ontology, psychology (so conceived) was a pretty obvious mistake just waiting to happen – an essentially derivative science modeled on physics, yet having as its subject the very realm that physics rendered utterly obscure." Hume's mechanics of the mind was formulated in experiential terms, viz. with the notions of impressions (corresponding to the positivists' notion of "sense data"), ideas (thoughts) and the relations between them.

[8] We do not experience a self, Hume argued, and therefore there is no such thing: "For my part, when I enter most intimately into what I call *myself*, I always stumble on some particular perception or other, of heat and cold, light or shade, love or hatred, pain or pleasure. I never can catch *myself* at any time without a perception. When my perceptions are remov'd for any time, as by sound sleep; so long am I insensible of *myself*, and may truly be said not to exist" (Hume, 1978:252).

Atomism

Finally, Hume's psychologization of meanings and morals was connected to an atomistic view of society, positing individuals as primary societal atoms, creating communities through social contracts.

Hume's moral subjectivism, his focus on subjective experiences, and his social atomism together served as a theoretical backdrop to our psychological social imaginary, i.e., to the contemporary manner of conceiving social life in psychological terms. I shall argue that this way of thinking about social life is wrong. This assertion, however, presupposes that we can meaningfully talk about social imaginaries being "wrong." How so? Although social imaginaries are constitutive of practices and thereby of social life, I believe along with Taylor that they can sometimes distort and cover over certain realities (Taylor, 2004:183). The atomistic view of society, for example, clearly distorts and misrepresents social life by depicting individuals as primordially socially disembedded. In reality, we are always socially embedded, for we can only learn who we are by being inducted into a language, a set of practices and a form of life, all of which are irreducibly social. What we may learn in the process, however, is to *be* an individual.[9]

Rousseau: The Deep Interior

Rousseau's *Confessions*, written in 1770, but published only after his death, begin with the following declaration:

> I am commencing an undertaking, hitherto without precedent, and which will never find an imitator. I desire to set before my fellows the likeness of a man in all the truth of nature, and that man myself. Myself alone! I know the feelings of my heart, and I know men. I am not made like any of those I have seen; I venture to believe that I am not made like any of those who are in existence. If I am not better, at least I am different (Rousseau, 1996:3).

Rousseau's *Confessions* puts premium importance on the unique individual and mark a new era in the history of literature. The narrative genres of earlier medieval and renaissance literature typically employed canonical models and archetypes. No particular persons were portrayed in these stories. But with Rousseau emerged the modern autobiography; a genre that not only depicts a single person and his experiences, but does so from the person's own point of view. This has been called the quintessentially modern mode of life-narration (Taylor, 1989:289). It stands out from previous forms of literature in its representation of a particular life in great and intimate detail, which reflects a changed view of the person. The human being is no longer an element in the cosmic order, but a psychological *self* that can narrate

---

[9] Taylor (2004) distinguishes between a *formal* mode of social embedding (a level on which we are always socially embedded) and a *material* mode of social embedding (a level of content, where we may indeed learn to be individuals) (p. 65).

its own story. What comes into existence is the "disengaged, particular self, whose identity is constituted in memory" (p. 288).

It is obvious that Rousseau, with his *Confessions*, alluded to a book with an identical title written by the Christian monk St. Augustine circa 400 AD (Hartle, 1983). The differences between Augustine's and Rousseau's *Confessions* are remarkable and instructive: Augustine's autobiography tells the story of a man's journey towards God, whereas Rousseau's book is about a man's journey towards himself, towards his own psychological life, so to speak. While Rousseau began his *Confessions* with a praise of his own uniqueness and singularity, Augustine began his corresponding book with a praise of God, belittling himself:

> 'Great art thou, O Lord, and greatly to be praised; great is thy power, and infinite is thy wisdom.' And man desires to praise thee, for he is a part of thy creation; he bears his mortality about with him and carries the evidence of his sin and the proof that thou dost resist the proud. Still he desires to praise thee, this man who is only a small part of thy creation. Thou hast prompted him, that he should delight to praise thee, for thou hast made us for thyself and restless is our heart until it comes to rest in thee.[10]

The *theocentric* worldview of Augustine is in stark contrast to the modern *anthropocentric* worldview of Rousseau. Augustine is an important figure in the history of psychologization, because he represents the transition from the *ontic logos* of the Greeks and their concept of the world as a meaningful, ordered whole, to the meaningless universe depicted and explored by Newton and Hume. In the mechanical, meaningless universe, Rousseau had to turn inwards to the newly discovered inner self to find meaning and value. Already Augustine had initiated the modern preoccupation with the inner self, and he crystallized the will "as an independent discursive component of understanding," which is necessary "for being a self in the modern sense" (Bertelsen, 2002:749). However, Augustine merely saw the inner self as the *road* to God and salvation. The inner self was never conceived by him as representing God or salvation in itself, and the cosmos was still depicted as a meaningfully ordered external structure. With Augustine, we are still far from the modern psychological worldview of Hume and Rousseau, where the world became reduced to the perceptions of the mind – as in Hume – and where the inner self should consequently be protected from the corrupting influences of that which is outside – society – as in Rousseau.

In *Emile*, Rousseau gave a psychological solution to the problem of how to form the self so that the unfortunate influences of sociality could be avoided, and in *The Social Contract* he gave a corresponding political solution to this problem, pointing to the establishment of the right democratic order (Reath, 2001). Rousseau here presented his version of social contract theory that (just as in Hume's case) portrayed isolated individuals as only derivatively coming together to form a society. Like in Hume's case, Rousseau's (social and political) philosophy is built on his basic psychology. Psychology takes precedence over the normative questions.

Rousseau's *dictum* "back to nature" advocates a return to what he perceived as the self-sufficiency of the inner, private self in childhood (Hartle, 1983:6). The first

[10] Quoted from the internet edition of *The Confessions* at: http://www.ccel.org/a/augustine/confessions/confessions_enchiridion.txt.

sentence in *Emile* is: "God makes all things good; man meddles with them and they become evil" (Rousseau, 1762:5). The impulses of nature, hardwired in our original childhood self, are always good. In Rousseau's moral psychology, evil enters the world only with human societies. There is no original sin, and only culture and sociality deprave us:

> Let us lay down as an incontrovertible rule that the first impulses of nature are always right; there is no original sin in the human heart, the how and why of the entrance of every sin can be traced. The only natural passion is self-love or selfishness taken in a wider sense. This selfishness is good in itself and in relation to ourselves; and as the child has no necessary relations to other people he is naturally indifferent to them (Rousseau, 1762:56).

The goal of life – what would be called *self-realization* by twentieth century humanists – is to return to the natural and original self, and the means is to turn inwards. We should learn to listen to the inner voice that speaks in us, and this demands independence from the pressures of society (Hartle, 1983:156; Taylor, 1989:359).

The inner voice of nature speaks with moral authority, according to Rousseau. It does not merely *point* to what is good and worthwhile, but *defines* it (Taylor, 1989:357). Rousseau here gives modern moral subjectivism its language, although he did not take the subjectivist turn fully, for, as Taylor says: "He ran his inner voice in tandem with the traditional way of understanding and recognizing universal good" (p. 362). However, in spite of their differences, the accord with Hume is striking: Both introduced moral subjectivism by arguing that what is good is good because human beings *de facto* like it. And both thought that humans are naturally endowed with quite similar preferences: Rousseau stressed the capacity of the original self to determine the good, and Hume stressed humans' natural sympathy towards one another.

Rousseau's heirs are the modern self-realization psychologies, especially the humanistic *third force* psychologists, but also more broadly those numerous psychologists who work in therapy, education, and organizations to enhance human autonomy and self-development: Rousseau "is the starting point of a transformation in modern culture towards a deeper inwardness and a radical autonomy" (Taylor, 1989:363). Many everyday practices today are organized in accordance with an ethic of the self-realizing, autonomous self (Rose, 1996a:17). According to Rose, the ideal of autonomy creates "an intense and continuous self-scrutiny, self-dissatisfaction and self-striving to live our autonomous lives, to discover who we really are, to realize our potentials and shape our lifestyles," by which we become "tied to the project of our own identity and bound in new ways into the pedagogies of expertise" (Rose, 1999b:193). This is one consequence of imagining social life in psychological terms that will be further discussed in later chapters.

Ontologizing Methods: Hume in Modern Psychology

In what follows, I describe how Hume's modernist methodological imperative to introduce the experimental method into moral subjects has influenced scientific psychology. I shall focus on the technical apparatus of psychology – "the experimental methods of reasoning" in Hume's words – and how these have shaped our psychological social imaginary. I concentrate on three kinds of experimental methods of reasoning: the experiment, the mental test, and statistics, which I will describe very briefly. Then we shall see how the methods employed by psychologists have become ontologized, read into the constitution of mind itself, with the effect of excluding *bona fide* moral normativity in our theories about the mind.

The Psychological Experiment

The psychological experiment as first practiced in Germany from the middle of the nineteenth century was an institution quite specific in time and place. It was founded on a combination of certain philosophical ideas and physiological investigative practices. If one takes an ethnographic stance towards the psychological experiment, it comes to look like a quite curious social institution. Today, the institutional arrangement of a psychological experiment has become well known. Experiments are known to such an extent that we often forget the numerous things that must be taken for granted in order for the practices of the experimental institution to proceed smoothly. All participants must be willing to abide by the rules and conventions of the experiment (Danziger, 1990:9–10). It is no use if subjects begin to fumble with the technical equipment, if they begin questioning the experimental set-up, if they address the experimenter in too friendly a way and try to engage in chitchat or something like that. Today, the psychological experiment has become a common social institution in the West, not just in its strict scientific form, but also in derived forms and through popularization in the mass media.

This was not so before people began to imagine social life in psychological terms. Even Wundt's early experiments followed other rules than today's psychological experiments. In Wundt's experiments, there was a relatively symmetrical relationship between experimenter and subject, often with the subject having the leading role. Experimenter and subject also frequently changed roles, which would be unthinkable today.

> Wundt appeared regularly as a subject or data source in the experiments published by his students, although he also contributed much of the theory underlying these experiments. [...] The participants in these experiments clearly saw themselves as engaged in a common enterprise, in which all the participants were regarded as collaborators, including the person who happened to be functioning as the experimental subject at any particular time (Danziger, 1990:51).

In the years after Wundt, however, "experimenter and subject roles are less and less frequently exchanged and research subjects are less and less frequently identified by name" (Danziger, 1990:73). After Wundt, the subject gradually became de-personalized, the subject became everyone, and in many ways, everyone literally became a subject as experimental practices spread.

The Psychological Test

While the rise of the psychological experiment was largely a German affair, the rise of the psychological test largely took place in Great Britain. Experimentation and testing are in many ways opposed, yet complementary practices: experimentation seeks to "maximize the demonstration of manipulative effects" whereas mental testing seeks to "minimize such effects" (Danziger, 1996:25). Francis Galton was the main figure behind the institution of testing in Britain. In 1884 he charged every person who came to be tested ("measured") in his laboratory the sum of three pence, and more than 9,000 people showed up. But, as Danziger remarks, Galton's interest in devising his "antropometric measurement" was not financial, but how the data could be useful in his eugenics program (Danziger, 1990:56). Galton was one of the leading architects in the "scientific racism" of the nineteenth century (Richards, 1996:164), and he was very much interested in practical social planning. The Galtonian mental test gradually replaced the collaboratory Wundtian style of experimentation as dominant in psychology, and a probable reason is that testing was more readily applicable in a range of different societal practices (Danziger, 1990:118). And testing methods *were* applied on a large scale. They became part of school life in the form of scientifically based examinations (p. 109), and they entered clinics, factories, and the military (Rose, 1999a:Chap. 4).

A main point emerging from Danziger's history of the subject in psychology is that from the very beginning of the twentieth century, psychology became an applied science, and an extremely successful one, which became involved in the constitution of the subjects that it studied. It was thus primarily the applied aspects that led to the psychologization of society. Psychological practices did not spread because of a theoretical insight into what the mind is like. Rather, it was specific investigative methods – "experimental methods of reasoning" in Hume's words – that made everyone see her- or himself in psychology's image. It was the very methods in psychology – experiments and tests – which led psychologists to devise new models of human beings, which again became part of the self-understanding of these human beings (this is an example of "the looping effect" to be discussed in Chap. 4). The subjects came to see themselves in terms of psychologists' research methods. Psychology's methods were ontologized.

Such ontologization often happens when psychology identifies its measures with the objects investigated. The categories of stimulus and response represent an instructive example (Danziger, 1996:21). Stimulus and response are intelligible and common, as units of measurement in psychology, but a lot of work has to be done by psychologists in order to crystallize such units in experimental practices. Neither our phenomenological experiences nor our stream of behavior come neatly and automatically arranged into these units. They are not given to pick up in nature. Imagining and arranging human lives in terms of stimuli and responses demands a highly constricted experimental environment. But, Danziger remarks, "stimuli and responses were always discussed as though they were features of the objective world and not artifacts of psychological procedure" (p. 21). These units, produced and employed by psychologists, were then identified with the "ultimate building blocks of reality" (p. 21). And when human beings begin to interpret their own and others' behavior in light of what psychology tells them are the ultimate building blocks of psychological reality, then we have come full circle in the process whereby methods are ontologized.

Psychological Statistics

Modern psychology began its life with an object of investigation inherited from a certain cultural and philosophical tradition (Danziger, 1996) – Hume's worldview of the eighteenth century – and from there, psychology went on

> to apply certain procedures of experimentation and quantification to the study of the pre-existing object. But once the disciplinary apparatus of investigation had been institutionalized, the possibility emerged of allowing this apparatus, rather than tradition, to define the objects of psychological science (Danziger, 1996:22).

Often the procedures came to dictate the theoretical formulations rather than the other way around (Danziger, 1996). The clearest example of psychology having identified its methods with its objects – what I call ontologizing methods – is found in statistics. Statistics originally emerged, as testifies its name, as a "science of state" (Rose, 1996b:111), as a technology intended to gather information about the states' populations in order to govern them. Hacking (1990) has argued that in the nineteenth century, with the development of statistical tools (largely due to psychologists such as Galton and Spearman), the belief spread that statistical laws expressed real laws inherent in social life. Statistical laws were no longer understood as simply expressing underlying deterministic events, for "statistical regularity underlay the apparently disorderly variability of phenomena" (Rose, 1996b:112). Statistics were ontologized – the world itself was seen as ordered statistically.

This has also been analyzed by Gerd Gigerenzer (1996) in an investigation of how psychological discoveries are dependent on psychologists' methods of justification for their knowledge claims. Gigerenzer's analysis demonstrates that "Scientists' tools for justification provide the metaphors and concepts for their theories" (p. 36). "Discovery is therefore, inspired by justification" (p. 46).[11] In psychology, the role of statistical tools was very important in this regard: "After the institutionalization of inferential statistics, a broad range of cognitive processes, conscious and unconscious, elementary and complex, was reinterpreted as involving 'intuitive statistics'" (p. 39). Psychological theories of the mind were formulated with clear inspiration from the new methods and tools for data *analysis*, rather than from new data (p. 38). With the advent of statistics, the mind of the human being itself was being framed as a statistician.

---

[11] Sometimes the natural sciences also work like this: In astronomy, once the mechanical clock was invented, the universe itself quickly came to be understood as one such mechanical clock (Gigerenzer, 1996:37).

Already in the 1940s had Egon Brunswik claimed that people are intuitive statisticians (Smith, 1997:838). Later on, also the computer became an extremely important tool that inspired cognitive theories about the mind. Thereby, the algorithms and operations of the computer became ontologized. Also the view of humans as probabilistic rational choice machines – the *homo oeconomicus* – owed much to the invention of statistics. Seen in this light, psychologists' methodological tools are not neutral, because the mind is continually recreated in their image (Gigerenzer, 1996:55). And the statistical view in psychology has at times gained something like scientific hegemony. Danziger sums this up:

> The more rigidly the demands of a particular statistical methodology were enforced, the more effectively were ideas that did not fit the underlying model, removed from serious consideration. Such ideas had first to be translated into a theoretical language that conformed to the reigning model before they could be seriously considered. In other words, they had to be eviscerated to the point where they no longer constituted a threat to the dominant system of preconceptions guiding investigative practices. *The final stage of this process was reached, when the statistical models on which psychologists had based their own practice were duplicated in their theories about human cognition in general* (Danziger, 1990; my emphases, SB).

When it had become evident that the object of psychological research – the mind itself – works statistically, there was all the reason in the world to concentrate on this method when doing psychological science. Methods and theories then confirmed each other circularly. Already in 1955, more than 80% of published experimental articles in scientific journals used inferential statistics as a means of justification (Smith, 1997:838). The experimental method of reasoning had, in the form of statistics, been introduced deeply into moral subjects, who themselves were now portrayed as statisticians. My question is, however, how algorithmic, rational-choice machines, which operate statistically in order to reach their desired goals, can act as moral beings? The answer seems to be that they cannot; they can calculate the optimal way of reaching their goals, but they seem incapable of judging whether their goals are *worth* striving for. A mind described as a machine works *mechanically* and *causally*, but never *normatively* and *morally*.

Macro Ontologization

Charles Taylor's (1988; 1989) analyses of the Western history of the self also demonstrate, on a historical macro level, how methods and procedures became ontologized. With the breakdown of the Greek and medieval teleological worldview caused by such figures as Galileo,[12] Descartes was famously prompted to formulate a philosophy of the disengaged mind (see also Toulmin, 1990). The new natural sciences worked very successfully by disengaging humans from the natural world

[12] Galileo died in 1642 after having created the first consistent mathematical theory of motion, and having claimed that the book of Nature is written in the language of mathematics.

through scientific procedures, and this newly developed capacity for disengagement was exported to other fields of inquiry, and influenced the images of mind articulated by Descartes, Locke, Hume, Kant, and later the science of psychology. The disengaged perspective was thus ontologized (Taylor, 1995b:66). The West witnessed "a kind of ontologizing of rational procedure [...], what were seen as the proper procedures of the rational thought were read into the very constitution of the mind" (Taylor, 1993:317–318). The mind was identified with a rational procedure. It was thus ignored that "Psychological reflectivity is a historical and societal product" (Poulsen, 1995:5), a product instantiated in many respects by "The advent of psychology, helping people to acquire an increasingly mediated relation to their daily activities" (p. 17).

There is nothing universal about being a procedural, disengaged mind or an intuitive statistician. Subjects that function like this, as many of us have come to do today, are historical products. Furthermore, if the points of Danziger and Gigerenzer are valid, it appears that the very methods developed in psychology are at least partly responsible for the fact that we have become such disengaged subjects. This has happened because psychology has worked by identifying its measures with its objects investigated, and in turn because these objects – human beings – easily identify themselves with how they are represented in psychological theories.[13] In short, we have witnessed the realization of Hume's program of introducing the experimental method of reasoning into moral subjects. The result has been that the human capacity for moral action has been ignored at best, or reduced away at worst, for a mind that is recreated in the light of value-neutral methods can hardly see itself in moral terms. As Husserl argued in his critique of scientism: Pure factual sciences make pure factual men (Husserl, 1954:4).

Free to be One's Self: Rousseau in Modern Psychology

So far I have approached the modern psychological social imaginary in terms of the rise of the disciplinary apparatus of psychology. This is the Humean, modernist, and scientific story about the introduction of the experimental method of reasoning into moral subjects. There is an equally significant story about the rise of a psychotherapeutic ethos (Illouz, 2008). Concerning clinical psychology and psychotherapy, much of this is recounted in Philip Cushman's works (1990; 1995). Cushman explores how the psychological healing professions have used different technologies to create, shape, and maintain a historically specific human subject. He is deeply critical of psychotherapy as practiced in the US, and argues that social, moral, and political problems are persistently psychologized and individualized by

[13] Of course, this is rarely a conscious and voluntary process. Often, as Foucault has taught us, we identify with specific representations of ourselves only through processes of subjugation and domination. Much more on this in the following chapters.

the therapeutic profession. He depicts psychology as "one of the guilds most responsible for determining the proper way of being human [...], especially in our current era, in which the moral authority of most religious and philosophical institutions has been called into question" (Cushman, 1995:336). Cushman points to the practical and applied aspects of psychology as responsible for the psychologization of the world: "Through the activities of what was called 'applied psychology', psychology would be the social science perhaps most responsible for the continued dominance of self-contained individualism and the resurgence of capitalism" (p. 160). If we bring together Cushman with the analyses recounted above, a picture emerges of a psychological discipline that has been deeply involved in the constitution of its object – human subjects – particularly because of the consequences of its methodological and therapeutic technologies having penetrated our social imaginary and social practices. The focus on methodological technologies is in direct continuation of Hume's modernist project, while the focus on therapeutic technologies for self-exploration and self-development are in continuation of Rousseau's romanticism, as we shall now see.

The most important heirs of Rousseau's ideas about the inner self and its realization are the humanistic psychologists. Humanistic psychology was developed in the US in the years following World War II, and was in its own eyes an alternative to psychoanalysis, where the individual was understood as controlled by unconscious forced rooted in childhood experiences, and behaviorism, where the individual was seen as governed by its reinforcement history. In opposition to these theories, humanistic psychology claimed that the healthy individual was not controlled by anything other than his or her own self. Its goal was to teach people to be free, as Carl Rogers said (1967a). To be free means to become what one really is: "It is the experience of becoming a more autonomous, more spontaneous, more confident person. It is the experience of freedom to be one's self" (p. 47). The goal is to become an "architect of the self" (p. 47).

According to Rogers, freedom means discovering that meaning is created from the inside; from one's own self and one's own experiences. It is the discovery that we ought not to be tied to anything but our authentic self. Rogers saw not just meaning, but also morality, as coming from the inside. The child has a healthy and clear approach to values based on immediate organismic evaluation. What is valuable is what the organism likes (Rogers, 1967b:19). This is strikingly similar to Hume's subjectivism. When the child grows up, however, it will try to obtain love and acceptance from its surroundings, and hereby the child easily gives up the idea that the source of values is inner, and instead places the source in other people. What is valuable becomes what the parents like. The child then acquires "a basic *dis*trust for his own experiencing as a guide to his behavior" (p. 17). The natural and original self is replaced by a false self.

The techniques developed by humanistic psychologists, and particularly through Rogers' work as a therapist, are designed to offer a way back to the authentic childhood self and the basic trust in one's own evaluations: "The locus of evaluation is again established firmly within the person" (Rogers, 1967b:22). In the mature, self-realizing person, evaluation again becomes a process, which is "fluid, flexible, based on this particular moment, and the degree to which this moment is experienced as enhancing and actualizing. Values are not held rigidly, but are continually changing" (p. 21). Something is valuable only if it contributes to the self-realization of the individual: "the criterion of the valuing process is the degree to which the object of the experience actualizes the individual himself" (p. 23). The self-realizing person acknowledges this and trusts his or her own natural self, rather than the experience of others: "*evaluation by others is not a guide for me.* [...] *Experience is, for me, the highest authority.* The touchstone of validity is my own experience" (Rogers, 1961:23). Both Rousseau and humanistic psychologists thus describe the development of the individual from a natural and wholesome condition in childhood, where the inner self dictates what is good and bad, towards the development of a false self, which arises because of the corrupting influence of culture and sociality. In order to become who we *really* are, we ought therefore to learn to listen to the inner voice of our organismic evaluation. This is the process of self-realization, which has since become dispersed across Western societies (see Chap. 3).

Subjectivization of Morality

Like Rousseau, Rogers clearly conferred moral values to the inner self: "I am the one who determines the value of an experience for me" (1961:122). It is only the individual's own subjective evaluation, based on the inner experiences, that can give value to something: "the individual cannot borrow value, truth, and meaning from without, but must create them from within," as it was put in a review article of humanistic psychology (Urban, 1983:161). Any external source of value is considered a threat to the individual's autonomy. This form of humanism reduces morality to psychology. The value of anything is determined by its psychological function. An example can clearly demonstrate the subjectivism and atomism in Rogers's thinking: According to him, the healthy family is no duty-bound, supra-individual whole, but consists of free individuals, who let each other become what they essentially are: "the family circle tends in the direction of becoming a number of separate and unique persons with individual goals and values, but bound together by real feelings" (Rogers, 1961:327). Family ties are, just as other "interpersonal relations," instrumental for individual self-realization. People should only maintain a relationship as long as "it is an enhancing, growing experience for each person" (Rogers, 1970:10).

The view of morality found in self-realization psychology is not just a subjectivized view, which confers the source of values to the subjective self, but also an emotivization, since it is the individual's emotions that determine the moral quality of actions and events. As Rogers said: "doing what 'feels right' proves to be a competent and trustworthy guide to behavior which is truly satisfying" (quoted in Vitz, 1994:54). Only with my feelings can I know if something contributes to my self-realization, and thus is good and valuable. The goal of therapy, therefore, is to create a relation in which "I am my real feelings" (Rogers, 1961:37).

Conclusion: Modernity and Psychology

In this chapter I have tried to trace the origins of some of the ideas that have been influential in shaping modernity's psychological social imaginary. These ideas are articulated in similar ways by the two faces of psychology's Janus head, which, however, are conventionally seen as incompatible. In reality, they are only incompatible on a superficial level, and they have both contributed to the creation of a psychological social imaginary. This psychological social imaginary involves:

1. A focus on private experiences as subject matter. A crucial point is that "experience" in this sense as it meets the individual is conceived as value-neutral. It is the individual who subjectively adds values unto the world (this view will be criticized in the second half of the book).
2. The psychological social imaginary involves a psychologizing of morality that presents it as a subjective phenomenon. This was evident in both Hume's 'modernism' and Rousseau's 'romanticism,' and the consequence has been that psychology has seen normative morality as something purely subjective, unavailable for serious consideration: "the exclusion of ethics and esthetics from access by scientific reasoning led over time to the denial that ethical statements could contain meaningful content other than an emotive expression of personal preference" (Polkinghorne, 1989:30). The latest psychological theory to endorse this view is perhaps evolutionary psychology.
3. The psychological social imaginary also revolves around a form of social atomism that I have only addressed in passing. Both Hume and Rousseau psychologized not just morality but also politics, and understood society as instrumental for individual needs. Interestingly, both of them backed their social theories with psychological assumptions about human experience. Psychology takes precedence over social theory, something we also see today with the (academic and public) success of evolutionary psychology, which is rapidly becoming a preferred source of explanation of almost any human phenomenon.

In modernity a new vision of moral order evolved that differed significantly from pre-modern notions. Humans were no longer parts of larger wholes (a community, a society, a cosmos) that defined the normative direction of their lives, for meaning and value came to be seen as inner, psychological phenomena. The scientific disenchantment of the world necessitated an enchantment of the mind with the birth of psychology as a result. If we think of "modernity" in broad terms,[14] then we can think of psychology as a central array of practices and techniques that evolve *with* and contribute *to* processes of modernization. What was invented in modernity's new social settings was the individual (Dewey, 1976; MacIntyre, 1985a:61); an

---

[14] Taylor defines modernity as "that historically unprecedented amalgam of new practices and institutional forms (science, technology, industrial production, urbanization), of new ways of living (individualism, secularization, instrumental rationality), and of new forms of malaise (alienation, meaninglessness, a sense of impending social dissolution)" (Taylor, 2002:91).

individual whom the new psychological social imaginary portrayed as ontologically prior to any social embeddedness.

My opening questions in this chapter were: In what ways is it possible to think about morality from the perspective of a psychological worldview? How have we learned to think about morality in an age dominated by psychological modes of understanding? The answer has been that imagining social life in psychological terms tends in the direction of an understanding of individuals, each with his or her private experiential realm that serves as the ultimate moral authority. This makes it difficult to think of normative issues as dependent on what the world is like. Morality then becomes a psychological phenomenon, and its normativity fades away. The psychological social imaginary presents moral goals as emanating from subjective minds, either in terms of passions (Hume) or the inner voice (Rousseau).

Chapter 3
Changing Psychologies, Subjectivities, and Moralities

That the ontology of the human being is historical has been claimed for many years by various strands of philosophy and social science. The historical nature of human self-interpretation plays a key role in interpretative or hermeneutic theory (Taylor, 1989), which finds some of its most important roots in Heidegger's analytic of *Dasein* (Heidegger, 1927). That the development of humanity should be seen as a shifting series of interpretations is also claimed by poststructuralists, particularly those influenced by the works of Foucault (1993). Both of these modes of thought and their relevance for psychology will be addressed in greater detail in the Chap. 4. In addition, historical materialists have long argued that the human being is a historical creature, whose ideas are inevitably shaped by the prevailing economic conditions. Marx himself argued that "the history of industry and industry as it objectively exists is an open book of the human faculties, which can be sensuously apprehended" (Marx, 1964:163). He further found that "No psychology for which this book, i.e., the most tangible and accessible part of history, remains closed can become a real science with a genuine content" (p. 164).

Psychology as a discipline has until recently kept this book closed by largely ignoring the historical nature of its subject matter, and even more so the historical nature of the discipline itself. Although there have been thoughtful arguments that the subject matter of psychology is historically constituted (Gergen, 1973, is a locus classicus here), the large majority of psychology suffers from historical blindness. Notable exceptions in this regard are the groundbreaking works of Danziger (1990, 1997a), Rose (1996a, 1999a), and Richards (1996, 2002), but, as the latter of these has put it, this new historiography of psychology remains marginalized in psychology at large (Richards, 2002:8).

The aim of this chapter is to outline some significant historical changes in the dominant ways that subjectivities and psychologies have been conceptualized and enacted, with significant consequences for our moral orders. The chapter outlines three ideal types of subjectivity that follows the well-known tripartite historical sequence of premodern, modern, and postmodern. One type of subjectivity is located in a premodern culture of *character*, another one in a modern culture of *personality*, and a final one in a postmodern culture of *identity*. Significantly, these cultures embody different moral outlooks. They embody different views of the

S. Brinkmann, *Psychology as a Moral Science: Perspectives on Normativity*,
DOI 10.1007/978-1-4419-7067-1_3, © Springer Science+Business Media, LLC 2011

relations between subjectivity and moral demands. I use the term "subjectivity" in a broad and generic sense to refer to our ways of interpreting what it is to be an acting and suffering human being. Subjectivity, in this sense, is a term I use to point to fundamental modes of human self-interpretation.

The analyses below follow the hermeneutic ideas that were sketched in Chap. 2, notably that human beings are self-interpreting creatures. This means that subjects are not discrete, given objects with properties that can be ascertained objectively from an outsider's perspective, but rather are dependent on how individuals interpret themselves. That the human being is a self-interpreting creature means, according to Charles Taylor, "that he cannot be understood simply as an object among objects, for his life incorporates an interpretation, an expression of what cannot exist unexpressed, because the self that is to be interpreted is essentially that of a being who self-interprets" (Taylor, 1985b:75). Our self-interpretations are not idiosyncratic mental representations or freely floating in some social ether, but are, according to Taylor's practice-oriented hermeneuticism, embedded in society's social practices. Our self-interpretations are produced by the spread of these practices and they also shape the practices and help them get established (Taylor, 2004:63). There is a dialectical relation between self-interpretations and social practices, just as there is between practices and society at large.

Our individual self-interpretations – for example the understandings of humans as being connected to the moral order through their characters, personalities, or identities – derive their contents and legitimacy from the practices of society and from what I previously called the social imaginary. The social imaginary affords certain self-interpretations, and precludes others. The social imaginary and its self-interpretations are dependent not just on vocabularies and symbolic discourses, but equally so on material practices and technologies. The technological aspects of subject formation have been stressed by Langman (2003), whose short history of subjectivity is cast in terms of the development of technologies of information. Premodern societies, he argues, created "subjects" (in the original sense) who were depending on the abilities of the powerful elite to interpret the world religiously by knowing specific privileged texts such as the Bible. Subjects should be loyal and have firm characters. In later modern cultures of literacy, individuals should think of themselves as citizens, and the nation states established standardized mass education, notably with the help of educational psychology and its technologies (e.g., the mental test). People in these modern, industrialized societies should develop stable selves in order to be able to enter and serve the workforce. Finally, dominant technologies today are the electronic media, giving rise to pervasive network and entertainment industries that afford self-interpretations of people as consumers with flexible identities.

A main point in what follows is that the history of humanity involves a development of the normative interpretive resources that humans have available for understanding themselves. It is this development that I shall attempt to lay out by focusing on three ideal types: Character, personality, and identity. In premodern times, certain practices dominated, which afforded self-interpretations centered on a concept of character. Many of these practices were religious. The question that framed people's subjectivities was "Who are you?" – addressing God as the great 'Other'. In order to understand themselves as subjects and interpret their lives, people had to understand the purposes and meanings that God had cast unto the universe. Life was lived in some *ontic logos* (Taylor, 1989) that defined the order, morality, and direction of a meaningful life. Later with modernity, the question in the culture of personality became: "*What* am I?", following the birth of the individual as an agent with discrete, measurable properties. One could now be understood, and understand oneself, as an individual in relative isolation from larger social (and cosmological) contexts. New practices concerned with measurements of individuals' characteristics, and also a romantic ethos of individualist expressionism (Berlin, 1999), supported the personality-oriented self-interpretation.

Finally, in a postmodern, consumerist culture of identity, the central question of subjectivity becomes "*Who* am I?", concerned not so much with discrete personality characteristics as with finding one's place in changing communities and subcultures. To the extent that we[1] today live what Zygmunt Bauman calls "the consuming life," we live a life that "is not about acquiring and possessing. [...] It is instead, first and foremost, about *being on the move*" (Bauman, 2007:98). Living with a feeling of being on the move makes identity and moral commitments a problem: How can I be *someone* when I am constantly changing? Thus, an explosion in identity discourse results when identities become problematic, as they did in the emerging consumer society in the post-war era. Consequently, as Stuart Hall has pointed out, there has been "a veritable discursive explosion in recent years around the concept of 'identity', at the same moment as it has been subjected to a searching critique" (Hall, 1996:1). Identity, in short, has become a central problematic for human subjects and their relation to the moral order today, in a way that character and personality were in earlier times.

1968 stands out as the (perhaps mythological) year of the emancipation of identities, witnessing both the many cultural revolutions that were to set identities free across the West and also the publication of Erikson's (1968) seminal *Identity: Youth and Crisis*. Unlike "personality," which is normally used to designate *what* somebody is (e.g., introvert–extrovert), "identity" points to *who* somebody is, and the postmodern condition is among other things characterized by a lack of stable markers of people's "who-ness." Hence we get the discussions that have filled social science journals for 20 years: Are identities social constructions or inner, mental structures? Are they constituted by personal experience or narrative

[1] When I say "we" here, I speak of a limited portion of the world's population, most of whom are located in the imagined hemisphere we call the West. But the self-interpretations of this group are affecting other parts of the globe, although not in a unidirectional way that meets no resistance. Opposition to "the consuming life" notably comes from different religious groups, and also from ecological and various anti-globalization groups in the West. No doubt, many people around the world interpret their lives within the frameworks of what I call "character" and "personality," but "identity" also plays a key role, e.g., for people who are active in anti-consumerist movements. My point is not that consumer society necessarily leads to the disappearance of character or personality, but that it centrally makes identity a problem that many people have to face.

conventions? Are they unities or multiplicities? Are they constructed within or outside discourse (Hall, 1996)? My aim in this chapter is not to answer these questions (I describe my own view on identity in Chap. 6), but rather to throw light on the developments that have led us to ask them and consider them important.

Although I shall argue that we now live in what is increasingly becoming a psychologized culture of identity, two points should be noted: First, that other self-interpretations are still important, and second, as Qvortrup (2003) has argued, that all historical epochs are "observed" rather than "existing." They are to some extent in the eyes of the beholder, and are often invoked to support someone's interests in the present. Qvortrup also argues, however, that this does not render macrohistorical characterizations useless, and he indeed builds his own argument around a three-stage view of history with the periods of traditional society (largely corresponding to my culture of character), modern society (like my culture of personality), and the hypercomplex society (much like my culture of identity). Such "epochalizations" are potentially useful as heuristic devices that can enable us to observe patterns, but we should be aware that they are heuristics rather than objective realities.

A Premodern Culture of Character

Concepts and ideas often appear in history when they are useful to certain groups or institutions. The concept of character was useful in cultures that embodied warrior ethics, e.g., in the ancient Greek civilization, and it is probably no coincidence that Greek moral philosophy was framed as *virtue ethics*. This was when reflective thought on character seems to have begun. As MacIntyre (2001) reminds us, originally in Greek philosophy, all ethics was virtue ethics, since it was concerned with the proper development of a virtuous character. The word "ethics" comes from *ethos*, the Greek word for "character." Determining the proper form of a virtuous character was not something for individuals to subjectively decide. Rather, it was determined by the cosmic order of eternal and unchanging ideas, as in Plato, or the immanent species-specific essence of a human being, as in Aristotle.

The Latin "moralis" was Cicero's translation of the Greek *ethikos*, which originally meant "pertaining to character" (MacIntyre, 1985a:38). A person's character was the disposition to behave systematically in a certain way, and moral philosophy in Aristotle's (1976) sense was a kind of characterology. To be "good" meant to perform one's natural function well. Humans were understood as entities with a proper function and purpose, and just as there is a difference between a good farmer and a bad farmer, there was an analogous difference between a good human being and a bad human being (this is taken up again in this book's part II). This difference concerned human character. One was a bad human being if one was defective in human *ethos* or character, and the normativity involved was seen as stemming from human nature itself. Ethics was the practical science of how to transform "man-as-he-happens-to-be" to "man-as-he-could-be-if-he-realized-his-essential-nature" (MacIntyre, 1985a:52). The essential means involved in such transformation were practical education and character formation through participation in the practices of the community, embodying a whole cosmology laden with intrinsic purposes.

Greek culture was thus a culture of character, at least for the aristocracy ruling the *polis*, where an overarching *ontic logo* determined how humans ought to live and develop. Ethics and the goal of life were predetermined by nature, and one's character simply ought to develop in a way that could help one achieve the preset goal. A related, but also markedly different, cultural ethos developed with the Christian faith, which also embodied a culture of character. Rieff (1965) and later Bauman (1996) have aptly captured the premodern and non-psychological conception of subjectivity with the ideal type of the pilgrim. The pilgrim travels towards a fixed goal, which is endowed with meaning because of the cosmic order or the community's practices. Certain shared ideals determine the direction of the journey through life. If one does not have sufficient character to live up to these ideals, a cure could involve different kinds of moral therapy, which were practiced in accordance with an explicit normative agenda (Lilleleht, 2003). Moral treatment was an essential premodern approach to the formation and correction of human subjects that later came to be associated with people like Tuke in England and Pinel in France (strictly speaking, this is hardly "premodern times," but their practices still incarnated the ethos of a culture of character). Moral therapy was a methodically concentrated use of premodern methods to instigate human improvement that had originally been an exclusive task for the church. Psychotherapy proper did not emerge as such until Freud developed his psychoanalysis, and with Freud we witness a change from a culture of character to a culture of personality (Susman, 1979).

A Modern Culture of Personality

Unlike moral therapy, Freud's psychoanalysis did not represent a morally authoritative perspective on a subject's character, but was a pure analytic approach to understand and treat the individual's personality. It has been claimed that psychoanalysis emerged as a tool to help people cope with living in a world of fragile social ties (Bellah, Madsen, Sullivan, & Tipton, 1985), and premodern moral therapies clearly functioned differently by tying people more firmly to the shared moral order that was already in place. Rieff has thus rightly described premodern moral therapy as "commitment therapy" (Rieff, 1965:68). The patient had to become committed to the community's symbolic system and the reigning moral demands. Like a pilgrim, the patient should develop a firm grasp of the undisputed goals (salvation, realization of the human essence), and developing one's character was closely related to developing a sense of one's duty vis-à-vis these goals.

The historian of psychology Thomas Leahey has described the moral essence of the concept of character and highlighted its contrast with personality: "Character was good or bad; personality was famous or infamous" (Leahey, 1994:246). With the emergence of modernity, the character related idea of doing one's duty remained important, but the notion of personality became increasingly central, not least because of the new science of psychology. In the premodern social imaginary, the subject was cosmologically decentered, and the existential task was to develop one's character in a way that would enable one to attain the largely unquestioned goal of existence, given by God or Nature. This whole worldview was supported by a teleological cosmology, and it was very much the later scientific revolution and its removal of meaning, value, and purpose from nature that led to the idea that it had to be individuals, human beings, who endowed the world with these qualities. These qualities could then be seen as things invented by humans, rather than things found in nature.

Romanticism and modernism, which were addressed in the Chap. 2 in the guises of Rousseau and Hume, are the two important sources behind the culture of personality. And although Rousseau's role in romanticism has probably been exaggerated, he must still be considered one of its fathers (Berlin, 1999:7). The idea that each human being has a unique personality that must be *expressed* as fully as possible is perhaps the central idea of romanticism, and one that has survived to our days. With Rousseau, as we saw, a new kind of self-interpretation and a new idea of social order were formed, framing the individual as a self-relying unit (in epistemology, this conception had already been inaugurated with Descartes). This paved the way for the modern preoccupation with personality. The other great source behind the culture of personality was modernism, i.e., the scientific ambition of knowing, measuring, and possibly improving the properties of individuals, not least their personalities. Originally, the concept of personality belonged to theology, law, and ethics, where it designated the moral aspects of the individual (much like "character"). Thus, when Kant used the concept in the eighteenth century, he used it to refer to the autonomous, ethical subject (Danziger, 1997a:124). But it has since lost its moral connotations and came to represent a distinct "scientific" view of subjectivity, and in the nineteenth century, the concept became gradually "medicalized" and Théodule Ribot could in 1885 for the first time write a treatise on *The Diseases of Personality*. It was now possible to have such diseases in analogy with somatic diseases, and this paved the way for the later interest in personality disorders.

According to Smith (1997:599), the modern psychological notion of personality is inseparable from the techniques that have historically been used to make every person measurable. The modernistic and scientific concept of personality and the related practices of measurement convinced many "that variations in intra-individual essence between individuals were purely quantitative, so that all individuals could be assigned a position on the same universal set of attributes" (Danziger, 1997a:128–129). Thus, a massive standardization of individual attributes followed with the introduction of "personality," and knowledge in this regard could be used instrumentally by people to achieve their goals: "Having a good personality demanded no conformity to moral order, but instead fulfilled the desires of the self and achieved power over others" (Leahey, 1994:246). With the modernist insistence on quantifiable properties of personalities, we are seemingly far away from Rousseau's romantic idea of the inner self, but Rousseau's way of imagining the individual, as a social atom with private attributes, is nonetheless an early expression of what later became a scientific preoccupation with measuring personality. Both the modernist and the romanticist stances were interested in *developing* and *realizing* the unique personality of the individual.

What effectuated the shift from character to personality in human self-interpretation? While character was what one should have in relation to a community's shared standards of correct and proper living (its moral order), personality became something to acquire and express when these standards were problematized with the transition to larger capitalist and state societies. In many ways, personality as a concept is connected to the more impersonal social order that was established with modern industrial society, embodying Durkheimian organic solidarity or a Tönniesian *Gesellschaft*. When the social order became impersonal, i.e., when one was no longer primarily defined by one's concrete relationships to significant others, the individual had to become "personal" in herself. That is, as a subject in the modern era, one must carry one's defining properties in and within oneself, and, in a sense, one has to acquire ownership over one's traits. When it becomes difficult to explain the actions of individuals with reference to their predetermined roles within specified groups ("he did that because he is the son of x"), people must look for explanatory clues elsewhere and personality here became relevant.

In accordance with the logic of standardization that was prevalent in industrial society, personalities also became standardized in the emerging culture of personality. At least in the sense that they came to be measured with reference to fixed parameters, tied to properties that persons allegedly have "in themselves," in relative independence of social surroundings. The reigning parameters are probably those that are laid out in the well-known Big Five models of personality (e.g., McCrae & Costa, 1987). People vary within the parameters, but the units of measurement are universal. A logic of standardization is found in industrial society at large, incarnated perhaps most clearly in Henry Ford's dictum that customers could have the Ford T in any color they wanted, as long as they wanted black (Paterson, 2006:61)! With the new methods of mass production, workers should often carry out one specific task or even movement during the workday. This logic of standardization characterizes not only modern industrial production, but also modern educational systems, bureaucracies, and the mentality of industrial society more generally. In modern industrial societies, the social imaginary conceived of the order as bureaucratized, technologized, and therefore regulated through larger structures than the closer and more personal social relations in groups that had previously established the social order in premodern times.

As an object for modern psychology, personality was from the outset constructed as something made up of individuals' measurable traits – personality became ontologized and conceived as something like a bureaucratic structure within the person – and an interest in measuring these traits came primarily from the industry and the military and the need for finding the proper person for the job. The first pen-and-paper test of personality was developed by Robert Woodwarth in 1917 with the intention of screening American recruits for participation in World War I (Ward, 2002:126). Later on, famous psychologists such as Ross Stagner, Henry Murray and Gordon Allport developed personality tests, and Allport also authored a very influential textbook on personality. The concept of personality was firmly institutionalized in psychology and many social practices outside universities in the 1930s in a way that the concept of identity was to become in the 1960s and 1970s. More recent personality psychology has continued the modernist quest for standardization, quantification, and value-neutrality. The Minnesota Multiphasic Personality Inventory (MMPI) was developed from 1943 onwards, and is a well-known example of a very widespread personality test that is used in clinics, hospitals, schools, organizations and prisons. Today, there are more than 2,500 publishers of personality tests that are also frequently used by recruitment companies.

It is clear that the culture of personality is not a monolithic form of self-interpretation. There are (at least) two significant strands involved, one romantic and one modernist, but both agree on a fundamental individualism and even atomism concerning individuals. The heirs to Rousseau's romanticism are primarily the modern self-realization psychologies, and the heirs to the modernist strand are those psychologists who are preoccupied with measuring this "thing" called personality, in organizations, schools, and personal life (e.g., in pop-psychological magazines). Both, however, often seem to de-moralize human subjectivity.

A Postmodern Culture of Identity

In latemodern or postmodern times, we have witnessed a number of significant changes in human self-interpretation, the social imaginary, and in how we conceive of the social and moral order. In premodern cosmologies, social order was thought to reflect an underlying natural order, given by God or simply the structure of the universe. In modern society, social order was understood as an achievement instantiated by rational individuals who were able to build structures, organizations, bureaucracies, and states (e.g., as articulated in the tradition of social contract philosophy of which Rousseau is a part). But in postmodern times, social order has come to be seen as much more fluid, changing, and contingent. Already Marx and Engels articulated this with the famous declaration in *The Communist Manifesto* that "all that is solid melts into air." They saw this as a result of a constant revolutionizing of production and disturbance of social conditions, which today seem more intense than ever. Some depict contemporary Western culture as a *network society*, which serves to highlight some of the ideological changes that have accompanied the recent historical transformations of capitalism.

Boltanski and Chiapello (2005) argue that the notion of network both as an organizational and institutional mechanism, and also as an ideology, has become central in postmodernity, characterized by flexibilization. In postmodern consumer society, the social imaginary leads people to imagine social existence as networks – the social simply *is* or has to a large extent *become* networks: "to the extent that the image of a society pervaded by networks takes hold as a fundamental means of societal self-description, other images of the social whole lose in influence" (Hartmann & Honneth, 2006:52). Other images still exist, but these are no longer leading. On a mundane level in people's lives, what used to be one's circle of friends and acquaintances is increasingly conceptualized as one's network that has to be continually monitored and adjusted according to one's entrepreneurial needs (e.g., through digital social networks such as Facebook). This also parallels many people's everyday experience in work organizations where the Weberian hierarchical pyramids of industrial society now become flattened and are organized as networks (Sennett, 1998). This is first and foremost a reaction to a market that has become much more fluid, flexible, and changing, and the way organizations and institutions naturally respond to this is by making themselves, fluid, flexible, and changing. Only networks, it is said, can adapt fast enough to the shifting preferences of consumers on a fast-changing, global market. Stability (in character or personality traits) is no longer a clear asset; flexibility, rather, is becoming important, including the ability to *adjust*, *develop*, and *learn* continuously (witnessed, for example, in the psychological discourse on "life-long learning"). If one is not "on the move," then one is "falling behind," and a new construction of human desire is involved: A desire to move by consuming rather than standing still by possessing: "new forms of consumption diminish possessiveness" (Sennett, 2006:8).

There are undoubtedly many contradictory trends and competing resources for self-interpretation today. Religion still plays an important role, especially in the United States, and it is unclear exactly how the current status of religious beliefs and practices is related to the emerging culture of identity and its moral order. Some parts of the religious landscape seem to be quite at odds with this culture and are actively opposing it (back to family values, a focus on character and stability), whereas other parts of the religious landscape thrive on network capitalism and offer new visions and practices of identity through self-improvement (which is reflected in much of the quasi-religious self-help literature, for example). In his recent grand investigation of secularism, Taylor concludes that what he calls "the expressivist revolution" – which followed "the new individual consumer culture released by post-war affluence" (Taylor, 2007:490) – has "undermined some of the large-scale religious forms of the Age of Mobilization" (p. 492). Certainly, a religious outlook on life is still possible, but religion is increasingly seen as connected to choice, authenticity, and, identity. In short, it is seen more and more as yet another lifestyle choice in accordance with the consumerist logic that is driving much of the culture of identity.

The dominant way of conceiving subjectivity both changes with, and has itself contributed to the change of the ongoing and intensifying transformation of the moral order. The logic of standardization behind the culture of personality is now supplemented with the logic of flexibilization inherent in the culture of postmodernism. This logic, I believe, also underlies the discursive explosion in identity talk. Sennett has pointed out that "As a general rule, identity concerns not so much what you do as where you belong" (Sennett, 2006:72). When one no longer automatically belongs somewhere, identity becomes a problem in people's lives, and social scientists and laypersons begin to address it. This has happened to a significant degree since 1968. Cushman and Gilford (1999) claim that we have now reached a point in consumer capitalism when it is no longer *efficient* to have a stable personality or an inner self to realize through prolonged training. The private self with discrete personality traits and properties is liquefied and transformed into heterogenous identities localized outside the person. The individual dissolves into situations – and is conceived as necessarily *situated* – its behaviors are no longer explained with reference to personality but to situational factors such as the desires that are invoked by commercials. This signals a shift from personality to identity discourse, and it gives us new moral challenges, for how can a liquefied subject be responsible and have long-term commitments across ever-changing situations?

Rosa (2003) has analyzed the transition from modern self-interpretations focusing on a stable, inner personality to postmodern self-interpretations oriented to "situational identities." He finds that the current self-interpretation of situational identity is expressed in the ways that people typically describe themselves. One no longer says "I am a carpenter," but rather "I work as a carpenter," because one wants to indicate that one is moving and possibly on the way to something new and better ("I live with Mary" rather than "I'm Mary's husband"). Rosa argues that such situational and temporary identity markers are a result of a general social acceleration, where everyone who is not moving, developing, and searching for one's identity is seen as standing still. The earlier strong feeling of direction in life is replaced by a complete lack of a sense of direction, and an almost compulsive tendency to be in continual movement, which, according to Rosa's analysis, is really a form of inertia (p. 20). It is thus no surprise that, typically, in social science literature, definitions of identity refer to fluid, situational factors. Etymologically, "identity" actually means "one and the same over time" (like when we talk about numerical identity: something being identical with itself), but, at least since Erikson, the concept has drifted into psychology and the social sciences in the sense of "belonging to a type or a group." This is almost the opposite meaning, and it is significant how most social science definitions of identity are built on the idea that identity concerns belonging (to a group), which is an idea that emerged exactly at the time when this form of belonging became a problem.[2] In the society of consumers, Bauman says, "no identities are gifts at birth, none is 'given', let alone once and for all in a secure fashion. Identities are projects: tasks yet to be undertaken" (Bauman, 2007:110). Also Erikson emphasized that identity is never something established, but is a never-ending process. It is not something one "possesses," but more like a continuous quest. Having an identity in the postmodern sense is closely related to what Bauman calls "consuming life," whereas the modernist preoccupation with personality was part of a culture of possessiveness.

[2] It is clear that identity means many different things in contemporary social science. In my view, a particularly constructive conception of identity links it with a person's value commitments (Taylor, 1989). In fact, this is also close to Erikson's original use of the term, and I develop this theme further in Chap. 6 (see also Brinkmann, 2008).

Changing Psychologies in the Transition from Industrial to Consumer Society

Along with the changing modes of human self-interpretation – from character to personality to identity – comes a change in how sciences and academic disciplines have approached human beings. Serious thought about character firmly belongs to moral philosophy and theology, and the birth of psychology is more or less coextensive with the transition to a culture of personality and the more recent culture of identity. The move from personality to identity goes hand in hand with a historical move from industrial society to consumer society, and I shall now examine more closely the psychologies that develop in these two historical contexts.

Kvale (2003) has argued that psychology in the modern industrialized world to some extent developed in response to the industry's need for qualified workers, and this was reflected most clearly not only in behaviorism, but also in psychoanalytic meta-theory, which was "replete with mechanical metaphors of the human psyche as a system of energy transformation" (p. 586). According to Kvale, psychology in industrial society could be understood in accordance with a factory metaphor. In contrast, psychology in the postmodern age can be understood in accordance with a market metaphor: "The visual and the symbolic landscapes are no longer dominated by churches or factories; today, it is shopping malls and all-pervasive advertisements that draw our attention" (p. 589).

In what follows, I briefly present four case studies of the psychologies of behaviorism, psychoanalysis, humanistic psychology, and social constructionism. I shall argue that there is not just a coincidental simultaneousness between the emergence of the two former ones and the zenith of industrial society, and the two latter ones and the emerging consumer culture in the years following World War II, but that these psychologies reflect, reinforce and – in some respects – can be said to have co-created the respective forms of capitalist societies and their moral orders.

Psychology in Industrial Society: Behaviorism and Psychoanalysis

Behaviorism and psychoanalysis are very different psychologies. Behaviorism emerged mainly in the US from the time when J.B. Watson wrote the behaviorist manifesto. Watson wanted psychology to be a "purely objective branch of natural science. Its theoretical goal is the prediction and control of behavior" (Watson, 1914:1). All this belongs to the well-known mythology of psychology, of course, and it has often been observed how Watson's main assumptions were in line with F.W. Taylor's scientific management (Kvale, 2003:587). More generally, it has been argued that the "mechanization of human behavior in industrial factories preceded the mechanization of behavior in psychological laboratories" (p. 588). From early on, psychologists sought to provide scientifically based advice to managers of factories, and an early and particularly clear example was Hugo

Münsterberg's *Psychology and Industrial Efficiency* from 1913. It was especially in the US that behaviorism and Taylorism had a strong practical impact and much less so in Britain (Thomson, 2006). Yet, it is fair to say that the behaviorist approach to psychology, stressing objectivity and mechanization, influenced deeply and globally how the mind was imagined not just by behaviorists proper, but also, for example, by later cognitive psychologists. Behaviorism's intention to establish psychology as a scientific discipline involved an exorcism of most of the phenomena that Freud's psychoanalysis, which had emerged a few years before Watson's manifesto, aimed to throw light upon. But, in spite of many differences, behaviorism and psychoanalysis can be seen as two sides of the same coin in industrial society. Following Freud's alleged claim (which in fact is nowhere found in his writings) that the healthy person can "love and work," we can simplify and say that behaviorism focused on the working side, whereas Freud himself focused on the loving side (psychoanalysis, however, was also applied directly in industry, for example by Elton Mayo, himself trained as a psychoanalyst, in the famous Hawthorne experiments that involved therapeutic interviews with workers; see Illouz, 2007:13).

Freud's psychoanalysis can be read as a description of the difficulties human beings confront in managing their drives and desires in a society that did not allow people much leeway in expressing themselves. Freud described the pathological conditions that ensued when people had too much drive and desire in relation to the moral order. Thus, in industrial society, people could develop neuroses, when their desires could not be contained by the ruling Victorian moral codes. Freud believed that different defense mechanisms were necessary in order to tighten the sexual reins, and such mechanisms could develop into pathological expressions, in extreme conditions in the form of hysterical blindness or paralysis. These are rare varieties of suffering today, which demonstrates that human psychopathologies are at least partly historically constituted (Hacking, 1995a). Freud described the neurosis as a symptom of a contradiction between societal suppression and sublimation of individual drives, and, consequently, this neurosis became the central social pathology (Honneth, 1996) of the industrial society. It prevented people from loving and working in satisfying ways. The neurosis was a symptom that the individual "wanted too much" in a culture that could not tolerate this due to its moral order and demand for stable selves. One may compare images of assembly lines and classrooms in schools from industrial society, both of which were spaces that clearly demanded strict discipline and control of human energy.

In contrast to industrial society, contemporary consumer society sees no problem in the idea that humans desire and are enterprising creatures. Today, it is no longer the surplus of energy and drives that represents a problem, but the shortfall of desire. People who lack the energy, drive, and desire to keep up with the pervasive demands to be flexible and interested in change and (self) development, are today visible as having a problem, which, in the most extreme cases, are conceptualized as pathological. The category of psychopathologies that represent an emptiness of energy is depression. The problem is no longer the management of drives ("I want too much!"). Today, instead, the central problem is a lack of energy ("I will never catch up!"). Stress has consequently been transformed from a physiological to a mental phenomenon (Viner, 1999), and psychologists have examined the now obvious links between stress and depression. In short, depression has become the central social pathology of consumer society. We live in "the age of depression" (Horwitz & Wakefield, 2005), for the person suffering from depression is visible as someone who does not correspond to the societal demands for flexibility, mobility, and a willingness to change and develop. The constant demand to find and realize one's true self can give people a sense of meaninglessness and emptiness (Cushman, 1990), and depression is a real risk when individuals are "psychically overburdened by the diffuse but widespread demand that they must be themselves" (Honneth, 2004:475). Honneth finds that this is a demand that perhaps was once emancipatory, but which now "has developed into an ideology and productive force of an economic system that is being deregulated: the expectations individuals had formed before they began to interpret their own lives as being an experimental process of self-discovery now recoil on them as demands issuing from without" (p. 474).

In summary, behaviorism provided society with technical solutions to the problems that arise when people fail to do what they are told, or fail to work with sufficient efficiency – in factories and in schools. Psychoanalysis provided society with more exciting and esthetically appealing solutions to the same kinds of problems in industrial society. Both psychologies reflected the moral order and addressed how to focus and manage human energy in a way that would secure a stable and ordered way of living. To a large extent, these psychologies fell out of fashion because the emerging consumer society no longer praised stability, but flexibility. After World War II, the goal became not just a transformation of neurotic suffering into ordinary suffering (following Freud), but to find oneself and realize one's true self.

Psychology in Consumer Society

Humanistic Psychology and Social Constructionism

In his study of the therapeutic construction of the self in American consumer society, Philip Cushman characterizes the role of humanistic psychology:

> Humanistic psychology's liberationist, transcendental, expressivist tendencies, combined with an optimistic pragmatic stance, moved in a direction often compatible with the energetic, flamboyant, on-the-make, sometimes nihilistic, always consumer-oriented post-war landscape (Cushman, 1995:243).

Humanistic *third force* psychology was developed by influential psychologists and psychotherapists such as Carl Rogers and Abraham Maslow in the US in the three decades after the Second World War. The other forces that humanistic psychology wanted to overthrow were behaviorism and psychoanalysis. Patients in the psychoanalytic practices were sick persons to be cured with the help of an expert. In humanistic psychology, patients became clients, and the client became the expert. As in consumer society at large, the client is always right. With its "promotion of spontaneity, of living out fantasies and desires, and with individual self-actualization as the goal of life," humanistic psychology came very close to "a consumer ideology" (Kvale, 2003:591). In the emerging consumer society in the years after World War II, referred to by Thomson (2006) as the permissive society, "members of Western societies were compelled, urged, or encouraged, for the sake of their own future, to place their very selves at the center of their own life-planning and practice" (Honneth, 2004:469). This became possible because of an extreme growth of income and leisure time, and consumer society thus provided fertile soil for humanistic psychology as an alternative to old-fashioned, pessimistic psychologies such as psychoanalysis. Humanistic psychology has since provided society with tools to shape itself and its members, not the least in work organizations through the Human Relations movement and later Human Resource Management, both significantly inspired by Rogers and Maslow (van Drunen, van Strien, & Haas, 2004). These practices sought to produce a flexible workforce, interested in personal development, which was and is in line with consumer society's need for flexible people who change and develop. In this sense, humanistic psychology was at once a product and a producer of consumer society.

Recently, perhaps the fastest growing field in psychology, known as positive psychology, has reactualized the ideas of humanistic psychology, albeit in a manner that is self-consciously more empirically strict than the research of the original humanists. Martin Seligman, the leading figure of positive psychology, characterizes humanistic psychology as alienated from conventional empirical science, but he clearly sees his own brand of positive psychology as continuing the vision of the humanists, although on a firmer scientific basis (indeed, the title of his book *Authentic Happiness: Using the New Positive Psychology to Realize Your Potential for Lasting Fulfillment* sounds like something from the humanistic psychologists) (Seligman, 2002). Positive psychology is not just a science of the positive – human happiness, virtue and positive development – but also an applied discipline, and, in Held's (2004) critical eyes, positive psychology is part of what she sees as consumer society's "tyranny of the positive attitude," which unites it with postmodern social constructionist therapies, which focus on positive renarrativization and appreciative inquiry, to which we now turn.

In the culture of identity, as we saw, identities are no longer attached to the private, inner self to the same extent as before. Instead, the individual becomes a network of relations. The main psychological theory that has addressed this is social constructionism, advocated with particular force by Kenneth Gergen. I am aware that social constructionism means many different things and is something of an umbrella term, so, in what follows, I shall concentrate on Gergen's version, which has been applied to a range of social practices, including therapy, organizational development, and education. As Gergen says, social constructionism functions both as meta-theory, as social theory and as societal practice (Gergen, 2001b:2). Gergen claims that the self is a social construction, that all meanings are socially negotiable, that any social change in principle is possible and that science is a form of "poetic activism" that changes the world by redescribing it (Gergen, 1999:117). Social constructionism portrays a world in which there are no fixed points, and it encourages human beings to change and be spontaneous and flexible in identities. We ought to be able to tell many different stories about ourselves instead of clinging to some truth about who we are (Gergen, 1994:249).

According to the historian Jackson Lears (1983), the idea that the self was a social construction first arose in the US with such authors as William James and George Herbert Mead when the therapeutic ethos of consumer society was beginning to emerge. The consumerist self, which is constituted by taking the attitude of the other towards oneself, was clearly articulated in Mead's theory (Kvale, 2003), as a forerunner of social constructionism. In general, social constructionism can be seen as a social science backdrop to the culture of consumption (cf. Brinkmann, 2006c).

As Sennett (1998) has argued, the new flexible and consumption-oriented capitalism threatens to engender a corrosion of character. Equivalently, social constructionism has been in the front criticizing dated Cartesian assumptions about an inner, stable and true self, and instead advocates change and spontaneity, flexibility in meanings and identities. In principle, according to this approach, we can always flexibly renegotiate meanings and construct a more appealing life narrative. Human beings are seen as networks of relationships, and it is interesting that one of Gergen's favorite images of his ideal of "relationship without self or community" or "pure relatedness" is the Internet (see Gergen, 2001b:194). Like positive psychology, social constructionism favors the positive over the tragic dimensions of life. In consumerism, there is a parallel hegemonic focus on entertainment, experiences and spectacles, rather than on existential vulnerability. Gergen himself publishes the *Positive Aging Newsletter* with the goal of socially reconstructing aging by presenting it as a time for further development and life-long learning. Another example is the constructionist method of organizational development called *appreciative inquiry*. This method works by directing people's attention away from "problems-talk" and toward positive reconstruction of identities and possible futures: "When organizations confront conflicts – between management and workers, men and women, blacks and whites and so on – appreciative inquiry shifts the focus away from problem talk [...]. Rather the attempt is to work with the organization to locate stories of desirable or ideal relations" (Gergen, 2001b:181). Constructionists regard all problems, meanings and values as *constructions* that can in principle always be redescribed to be made to look good.

The general tendency of signs and meanings in consumer societies to become disconnected from their worldly referents – what Lefebvre (1968) called "the decline of the referentials" – has become a central part of the social constructionist account of meaning: "our statements of belief contain words that are not fixed in their meaning [...] Everything that is said could be otherwise" (Gergen, 1999:161), "constructionists take meaning to be continuously negotiable" (p. 236). The ideal is to be able to perform one's identity in a flexible manner that will secure that "the dialogue" goes on. For constructionists, everything is negotiable, everything could be otherwise, and we all need the hermeneutic skills that enable us to make things look one way or the other. Accordingly, in a consumer economy, money is the universal hermeneutic medium that can transform anything into anything else.

In consumer society, products and life styles are judged according to esthetic standards. The question about how to live one's life can be answered, not just in religious or ethical terms, but increasingly also in pure esthetic terms (cf. Bauman on the tourist as a life ideal). This is crystallized in Foucault's well-known remark how, after having realized today that the self is not something given, we have to create ourselves as a work of art (Foucault, 1984). Gergen's social constructionism follows Foucault's "aesthetics of existence." Consequently, researchers should not be political activists, nor moral activists, but poetic activists (Gergen, 1999:49): Poetic activism "asks us to take risks with words, shake up the conventions, generate new formations of intelligibility, new images, and sensitivities" (p. 117). The goal is destabilization and flexibilization. I shall discuss the deep moral problems of this in Chap. 6.

Concluding Comments

In this chapter, I first presented three historical cultures that differed concerning how they framed and enacted human subjectivity and self-interpretation. I called them a culture of character, of personality, and of identity, respectively. In a culture of character, one is looked upon (and looks upon oneself) in a moral and religious perspective; in a culture of personality, one is looked upon in a psychological (scientific as well as romantic) perspective; and in a culture of identity, one is looked upon in a consumerist perspective. The analysis has primarily been descriptive in the hope that the historical account would be convincing, but I have also tried to frame the descriptions within a hermeneutic theory of human beings as self-interpreting creatures. This is a theory according to which we should think of human self-understanding as arising out of a social imaginary, the historically established practices that we participate in, and constantly recreate and develop.

I also looked more closely at how psychology developed with the transition from industrial society to consumer society. If we return to the conception of the social imaginary, we can summarize two competing facets of what I called a psychologization of the social imaginary. In industrial society, psychology developed technologies for stabilizing selves, workers and learners. The stable self should function in hierarchical and bureaucratic structures, and a central problem for this kind of self was how to control desire. Behaviorism and psychoanalysis engaged with this problem in different ways. The former through devising laws of reinforcement that could be applied in schools, factories and child rearing, and the latter through a transformation of neurotic suffering into tolerable suffering, enabling men and women to love and work. The social was imagined as dualistic, e.g., involving a Habermasian dualistic theme of System vs. Lifeworld, with the System constantly threatening to colonize the Lifeworld. Aldous Huxley's *Brave New World* from 1932 illustrates the image of psychologists in the role of educators, conditioning individuals from their conception to death on behalf of the World State (whose telling motto in the book is "Community, Identity, Stability"), which represents an extreme psychological colonization of the Lifeworld.

In contemporary consumer society, the dualism of Lifeworld and System erodes (theoretically and practically), and if there is colonization, it is more likely the Lifeworld that colonizes the System. Formal bureaucracies become informal and fluid networks, the economic sphere becomes saturated with emotions (Illouz, 2007:23) and public life becomes saturated with intimacy (Sennett, 1977). Formal and static qualifications become renewable competences to be developed in a never-ending process of life-long learning. Identity moves away from traditional categories and toward "commercial categories associated with brands and consumables" (Barber, 2007:200). The central human problem is no longer to control desire, but to make choices concerning identity and life plans.

Humanistic psychology was born in the emerging consumer society, and social constructionism today radicalizes the idea of fluid, flexible, and networking selves. The humanistic idea of humans having an inner self to be developed still exerts some influence on how we imagine social life, but the constructionist ideal of "pure relatedness" has come to the fore in the latest phase of consumer capitalism. The "spirit of capitalism" has changed from "industrial assembly-line production combined with social engineering" in the early twentieth century, through "post-industrial restructuring," partly facilitated by the humanistic Human Relations movement in mid twentieth century, and finally into "postmodern flexibilization facilitated by social networking" in the late twentieth century (Kemple, 2007:152). Consequently, the contemporary social imaginary recasts the social as networks. Human desire is gradually transformed from stable possessiveness to flexible consumption. Psychology's role is no longer primarily to develop technologies for stabilizing selves, but rather to contribute to a flexibilization of the consumerist self. Many new psychological technologies such as coaching, appreciative inquiry, short-term therapy, stress management, and also quasi-esoteric practices like mindfulness therapy (which has also found its way into mainstream cognitive therapy) point to a psychological flexibilization of human beings. The downside may be a rising frequency of depression, which indicates that not everyone can catch up and be flexible.

The social imaginary refers to our taken for granted practices and background understandings. The point of this chapter has been to demonstrate how these have been psychologized in different ways in industrial and consumer societies. Certain practices focusing on stability, ownership, and control thus came to be seen as natural in the social imaginary of industrial society. Some of these were developed by psychologists to assist in organizing people's everyday lives around these themes. In consumer society, practices focusing on flexibility, consumption and change have become central to the moral order. Individuals should want to be energetic lifelong learners, flexible and networking selves, if they wish to be an asset in contemporary culture. Humanistic psychology assisted in dissolving the political into the personal, and psychological social constructionism has dissolved both the social and the personal into relational networks. The social imaginary has changed from an understanding of the individual selves adapting to pre-given structures (cf. the psychologies of industrial society), to an understanding of pre-given selves that should be realized and thereby change the culture (humanism) and finally, with constructionism, to an emerging vision of a limitless network sociality (Wittel, 2001). Like all other social imaginaries, these consist of complex arrays of practices, ideas, and normativities, and my argument has been that many of these have come from different forms of psychology.

Chapter 4
How Psychology Makes Up People

As we have seen in the two previous chapters, the psychological ways of thinking about people have not just served as passive representations of human subjects, but have in fact deeply influenced how humans think and feel, and indeed influenced human subjectivity itself. Since the birth of psychology, humans have increasingly come to think about themselves in light of psychology's concepts and categories, and their lives have become dependent on psychological technologies such as tests and therapies. Psychology is in the business of "making up people," to use Ian Hacking's catchy phrase (1986). One problem arising from this is sometimes referred to as the reflexive problem: Psychology "is produced by, produces, and is an instance of, its own subject matter" (Richards, 1996:5). In this chapter, I approach the reflexive problem as a *moral* problem. As philosopher of science Isabelle Stengers has put it:

> Every scientific question, because it is a vector of becoming, involves a responsibility. "Who are you to be asking me this question?" "Who am I to be asking you this question?" These are the interrogations that the scientist, who knows the irreducible link between the production of knowledge and the production of existence, cannot escape (Stengers, 2000:148).

Knowledge producing psychologists are co-producers of human existence, which gives them a moral responsibility.

In what follows, I ask what it tells us about the nature of psychology's subject matter if the discipline can "make up people." Following Ian Hacking and Kurt Danziger, I argue that it tells us that "psychological objects" are human kinds (what these are will be explained below). Psychological objects (such as intelligence, learning, depression) are not naturally existing objects, for, unlike natural kinds, their existence depends on certain descriptions and discursive practices. Unlike natural kinds, human kinds interact with their descriptions and categories; there is a "looping effect of human kinds" (Hacking, 1995b).

Next I discuss two competing ways of understanding "the looping effect of human kinds": How do the processes of "making up people" occur? First I give an account of the Foucauldian approach to this matter, particularly as it has been developed by Nikolas Rose's "analytic of human technologies" stressing what he calls the *techne* of psychology. Foucault himself thought that psychology was "a dubious science" (Dreyfus & Rabinow, 1982:xx) that will never reach the level of Kuhnian normal science, but which is deeply and unavoidably enmeshed in political,

S. Brinkmann, *Psychology as a Moral Science: Perspectives on Normativity*,
DOI 10.1007/978-1-4419-7067-1_4, © Springer Science+Business Media, LLC 2011

social, and cultural functions in different ways. From early on, Foucault was aware of the normative dimensions of psychological knowledge, that is, of its ability to constitute and transform its subject matter. In an early interview from 1965 he said: "I don't think psychology can ever dissociate itself from a certain normative program [...] Every psychology is a pedagogy, all decipherment is therapeutics: you cannot know without transforming" (Foucault, 1998b:255).

I subsequently compare the Foucauldian approach to the interpretive-pragmatic understanding of looping effects that is advocated in this book, which frames humans as self-interpreting beings, and which, unlike Foucault, aims to preserve a space for genuine moral values and meanings in the practices through which psychology "makes us up." Instead of beginning with the *techne* of psychology and a problematization of practices (Stern, 2000), the interpretive-pragmatic approach insists on beginning with hermeneutic *articulation*, and it preserves a role of meanings and values in changing and improving practices (Taylor, 2000:123). For that reason I argue that the hermeneutic or interpretive-pragmatic approach is morally superior to the esthetization of human existence found in the (late) works of Foucault.

Natural Kinds and Human Kinds

I shall begin by asking whether psychological categories pick out natural kinds. Natural kinds are such things as tigers, water, or gold. They are "groups of naturally occurring phenomena that inherently resemble each other and differ crucially from other phenomena" (Danziger, 1999:80). The search for natural kinds is the search for the underlying essences of things. Natural kinds allow us to explain other more superficial properties of a class of things. Take water as an example. Water has certain observable properties: it is colorless, tasteless, has a specific boiling point and freezing point and so on. The underlying essential property that allows us to explain all these "superficial" properties is its molecular structure (H_2O). Scientific categories can be said to refer to natural kinds if the term for them were to remain applicable even if we were to discover that they possessed completely different properties than we had traditionally thought (Collin, 1990). For example, if humans somehow developed finer senses of taste, and found that water is not really tasteless, we would still call it water. The reason for this is that its underlying essence remains the same, although its "superficial" phenomenal properties appear differently to us.

The discussion of natural kinds has been significant in modern philosophy of language, challenging widespread views of linguistic meaning. Most philosophers and psychologists have traditionally thought that meanings are mental entities of some sort. Knowing the meaning of "water" would, on this traditional account, involve being in a certain psychological state; having a certain mental representation. Philosophers Saul Kripke and Hilary Putnam successfully challenged this view in the early 1970s. Putnam, whom I concentrate on here, asserted: "Cut the pie any way you like, 'meanings' just ain't in the head!" (Putnam, 1973:704). Putnam defended an externalist account of meaning on the grounds that the meaning of such terms as "water" could not be accounted for as concepts, or indeed mental entities of any kind. Instead their meanings had to be understood as rigidly fixed to the natural kind picked out by the term. The illustrative and now-famous science-fiction example given by Putnam is the twin-earth example. We should imagine that an expedition from our planet earth finds another planet far away from ours, which in every detail is just like our planet. They name it twin-earth. On twin-earth, there is a liquid, which in all immediate phenomenal respects is just like our water. People on twin-earth have exactly the same practices with this liquid as we have with water on our planet. They drink it, boil it and freeze it, just like we do. But when the expedition analyses the liquid chemically, it turns out that its molecular structure is not H_2O, but XYZ. The question then is: how should the compound be classified? Putnam argued that we would intuitively say, and rightly so, that the compound is not water, but merely something that looks just like it in all observable respects. Putnam thus argued that natural kind terms (e.g., "water") refer to natural kinds ("essences," e.g., "H_2O"), just as proper names refer to particular people, without the mediation of mentally represented meanings or definitions. These terms are rigidly fixed to their referents.

If we accept this analysis of natural kind terms, the next question to ask is whether natural kinds exist in psychology, or indeed in any social science. I will argue that in general the answer is no. As things stand, we have reason to believe that very few, if any, of the central terms in social science refer to natural kinds.[1] It would not be conceivable, to take a simple example, to think that we could discover that a hospital is really not an institution whose purpose is to treat illnesses, but something completely different, for example, a place for developing means for biological warfare. If it turns out that some hospital de facto functions like this, we will have to conclude, not that we did not know what hospitals were, but that this specific institution was not a hospital. If some radical Foucauldian could show that all hospitals in fact have worked single-mindedly to oppress rather than cure people, the right conclusion would be that *bona fide* hospitals have never existed, only institutions calling themselves "hospitals," but with different, "un-hospital-like" underlying rationales.

To cite an example more congenial to psychology: It seems unlikely that we would say, even if we some day succeed in identifying depression with a specific neural dysfunction, that we would thereby have discovered a natural kind (the essence of depression). Because, to reiterate the twin-earth example, if it turns out that depressed people on twin-earth, that is, people with anhedonia and all other phenomenal expressions of depression, have a completely different neuronal make-up compared

[1]Of course, the social scientific disciplines make use of natural kind terms when they draw on insights from chemistry or biology, for example. But these, I would argue, are auxiliary terms rather than central ones. For example, psychologists sometimes invoke our genes (a natural kind term) to explain some psychological phenomenon (e.g., aggression), but the psychological phenomenon itself is not a natural kind. Or so I argue here.

to depressed people on earth, then we would still not give up our classification of these people as depressed. They would still be rightly described as "depressed" even though their neuronal structure corresponded to the neuronal structure of happy and carefree people on earth. Unlike water, there is not in the case of depression some underlying essence, which fixes the meaning of the concept "depression." Instead, its meaning is tied to the "superficial," observable properties of depressed behavior and its significance in the human life world.[2] In the human sciences, concepts cannot be about some completely hidden essence, because, as Anthony Giddens has argued, concepts in the social sciences are concepts about concepts (Giddens, 1976). That is, they are concepts that somehow *must* hook up with the participants' own concepts (this argument was also forcefully made by Winch, 1963).

Psychological categories thus have to relate to the phenomena of the life world and with "folk psychologies." John Dewey used to distinguish between two psychologies, and he argued that in psychology, "all phenomena can be divided into the physiological and the social, and that when we have relegated elementary sensation and appetite to the former head, all that is left of our mental life, our beliefs, ideas and desires, falls within the scope of social psychology" (Dewey, 1917:54). I am willing to grant that *some* categories that are used in psychological research may in fact refer to natural kinds. It cannot be excluded that there are genuine natural kind terms in parts of what Dewey called "physiological psychology," but in "social psychology," where psychological phenomena are *meaningful* and relate to human action, they are much harder to find.

I should make one important reservation, however. It is fair to say that the idea that the human being is a self-interpreting animal (Taylor, 1985b), and that humans are such creatures that are able to respond to normativity, also represent natural kinds. The human "essence," in this sense, seems to be not only that we lack an essence, and that we constantly understand ourselves in light of historically evolves categories, but also that we essentially live in some normative structure (space of reasons). This, however, is a very "thin" essence. Hacking (2004:281) argues that "It is in the nature of a human being to have no intrinsic nature," and he takes this not as a rejection of innate potentialities for specific mental traits, but as a reminder that "our genetic essence is not our essence" (p. 287). An individual's genes may determine "the extreme limits of possibilities," but, as Hacking makes clear, "it is *choices* that create one's character, one's veritable essence, one's soul" (p. 287; emphasis added, SB). And new ways to choose, including choices of who one is, arise as a result of new ways of describing people, as we shall see.

---

[2] A counter argument here is that we can perhaps imagine a future scenario where we diagnose depression according to serotonin levels and *completely* ignored the patient's feelings and statements about herself (Hacking, 2001, imagines such a case). My reply is: yes, I think it is likely that we are moving in that direction (see also Rose, 2003, 2007). However, if things develop that way, I think the correct response will be to say that the word "depression" will have acquired a new meaning, rather than to say that we have found the essence (the natural kind) of what we now call depression.

If the large majority of psychological categories do not refer to hidden essences or natural kinds, then what do they refer to? Hacking answers that psychological categories refer to human kinds.[3] What are these? Initially, we can outline three ways in which human kinds differ from natural kinds:

Natural kinds:

- Intelligible outside discursive contexts
- Indifferent to the descriptions applied to them
- Categories and kinds are independent

Human kinds:

- Intelligible only within a discursive context
- Interact with the descriptions applied to them
- Categories and kinds emerge together

Hacking says that "the chief difference between natural and human kinds is that the human kinds often make sense only within a certain social context" (Hacking, 1995b:362). One can be a king only in a world where the institution of monarchy exists (water, on the other hand, was H_2O before the institution of chemistry). Human kinds indicate "kinds of people, their behavior, their condition, kinds of action, kinds of temperament or tendency, kinds of emotion, and kinds of experience" (pp. 351–352). Hacking has analyzed in varying detail "multiple personality disorder" (Hacking, 1995a), "fugue" (Hacking, 1998), "homosexuality" (Hacking, 1986), "criminal behavior" (Hacking, 2001), "suicide," "teen-age pregnancy," "adolescence," "child abuse," "autism" and "Hispanic" (Hacking, 1995b) as examples of human kinds.[4]

An important feature of human kinds is that they can exert effects on themselves (Martin & Sugarman, 2001). They are affected by their classifications and interact with their classifications, sometimes affecting the classifications themselves. Human kinds can even intentionally try to change how they are classified (e.g., when homosexuals objected to being categorized as pathological in the DSM system). This is the looping effect of human kinds: "People classified in a certain way tend to conform to or grow into the ways they are described; but they also evolve in their own ways, so that the classifications and descriptions have to be constantly revised" (Hacking, 1995a:21). And further: "Inventing or molding a new kind, a new classification, of people or of behavior may create new ways to be a person, new choices to make, for good or evil. There are new descriptions and hence new actions under a description" (p. 239).

[3] In more recent writings, Hacking has replaced the terminology of human and natural kinds with that of interactive and indifferent kinds, but I do not think that this terminological shift is important in the present context (see Hacking, 1999).

[4] Furthermore in psychology, Danziger (1997a) has explicitly referred to Hacking's use of "human kinds" to analyze such kinds as "intelligence," "behavior," "learning," "motivation," "personality" and "attitudes." Danziger has also referred to psychologically developed human kinds as "psychological objects" (Danziger, 2003).

Why do new kinds of description, for example those produced by psychologists – Hacking (1995b) refers to Freud as "the king of the loopers" – provide for new kinds of action? Because in so far as human actions are *intentional*, they are actions under a description (Hacking, 1995a:234). We only say that people act intentionally if there is some kind of description of their action that renders it meaningful by locating it in a discursive context. I can only have the intention to vote for a given candidate at a democratic election if I can describe the physical behavior commonly associated with voting (going to the voting booth and marking the piece of paper) as an act of voting. One cannot act intentionally under a description unless this description is discursively available. As Hacking says:

> When new descriptions become available, when they come into circulation, or even when they become the sorts of things that it is all right to say, to think, then there are new things to choose to do. When new intentions become open to me, because new descriptions, new concepts, become available to me, I live in a world of new opportunities (Hacking, 1995a:236).

Human kinds are not unreal. As Hacking makes clear, some people do suffer from Multiple Personality Disorder, for example, although this is a human kind. They (and the communities in which they live) are not faking it, although they could not have acted the way they do, and *eo ipso* suffered from this condition, before the relevant field of description became available sometime late in the nineteenth century. Or – to be precise – late in the afternoon of July 27, 1885, when Charcot's student, Jules Voisin, described the very first MPD-patient, Louis Vivet (Hacking, 1995a:171). In contrast, it can reasonably be argued that people were able to suffer from tuberculosis even before this illness was named (although they may have been unaware of the cause of their disease). In the case of tuberculosis there is a relevant natural kind involved that causes the disease, which is not dependent on a discursive context, that is, on how it is described and classified.

I am aware that that point has been denied, most forcefully by Bruno Latour, who has argued that although we have found traces of (what we call) tuberculosis in the mummy of Pharaoh Ramses II, he could not have died of tuberculosis for the simple reason that the bacillus was not identified until more than 3,000 years after his death (Latour, 2000). Latour believes not only that all our categories are socially constructed (which is obviously true), but also that all kinds referred to, by our categories, including natural ones, are equally so constructed. All kinds are, to use Hacking's terminology, human ones, because all kinds interact with their classifications. Against Latour's monism of kinds, I want to argue that his view makes it difficult to explain why natural things resist our dealings with them in a way that human things do not necessarily do. Latour has fruitfully defined the real as that which resists us: "Whatever resist trial is real [...] The real is not one thing among others but rather a gradient of resistance" (Latour, 1988:158–159). (I advance a similar idea of "the real" in Chap. 6, when I argue that moral issues are real in the sense that we are forced to deal with them.) Latour is thus clear that we cannot socially construct nature as we would like it to be, but how can this be maintained if one is unwilling to grant that some parts of reality exist in some definite form external to social practice? If something resists our practices of classification, it might be because there are structures (natural kinds) prior to our categories that limit how we can fruitfully classify it.

Furthermore, unlike water and other natural kinds, human kinds are intrinsically part of social practice and take *an interest* in how they are classified. Human kinds are laden with values. This in fact is Hacking's argument against Latour's insistence that all natural kinds are at the bottom social (see Hacking, 1995b:365–368). It makes no difference to water what we call it: "Objects known about in the natural sciences do not change because they are classified, although we may change them in the light of our classifications" (Hacking, 2004:298). But it does make a crucial difference to humans how they are classified. And, after all, as Alasdair MacIntyre once remarked, molecules don't read chemistry textbooks, whereas humans do read psychology books that affect their self-understandings (MacIntyre, 1985b). Again, it is important to bear in mind that human kinds are just as real as natural kinds. There is no ontological difference between them in that sense. Human kinds do not belong to a different, "non-natural" world. In a sense they are just as natural as natural kinds, but they are natural in a different way, akin to what Hegel and modern Hegelians call "second nature" (Hegel, 1821:35; McDowell, 1994), that is, nature as it expresses itself in and through social practice. It is only on the assumption that humans are not part of nature that the dependence of human kinds upon human agents and practices could render them unreal (cf. the discussion of affordances in Costall, 2004b:84).

An Analytic of Human Technologies

I shall now consider two different frameworks that may help us understand practices of classification and "looping effects" in psychology: the Foucauldian and the interpretive-pragmatic. The goal is not to provide detailed analyses of actual looping effects, but rather to ask: What is the proper approach and attitude towards human kinds and their looping effects in psychology? What are the moral implications of the looping effects? Human kinds indicate something about human subjectivity. Human kind terms are always applied to, and by, human subjects. For that reason, the following discussion will center round the notion of subjectivity, and I begin with Foucault's conception of the subject, before moving on to Rose's "analytic of human technologies."

Foucault on the Subject

Foucault's historical and philosophical argument involves the view that the human subject as studied by the social sciences did not exist as an "object of truth" until early in the nineteenth century (Smith, 1997:857), when the social sciences began to create a subjugated subject by rendering it calculable, manageable, and governable.

Some of the technologies in use (experiments, tests, statistics, and therapies) were mentioned in previous chapters. Foucault's overarching goal had been "to create a history of the different modes by which, in our culture, human beings are made subjects" (Foucault, 1994b:326).

A radical point in *Discipline and Punish* (Foucault, 1977) is that the human subject is an effect of power. It is not primarily subjects, who, from a position outside power relations intentionally exercise power in order to promote their specific interests, for being a subject with interests in the first place is only possible because of power relations. Power is thus not merely oppressive, according to Foucault, but also productive: it produces subjectivities, as illustrated in his famous example: the *panopticon*. This was a prison structure developed by Jeremy Bentham, the father of utilitarian moral philosophy, which would enable a single guard to monitor the behaviors of a large number of prisoners. The *panopticon* should render individuals visible at all times, but should also make the individuals monitor themselves. Prisoners unavoidably turn the guards' gaze towards themselves, and thereby become constituted as self-monitoring or self-examining subjects. A self-examining individual has the soul, psyche, or subjectivity as a correlate (Coles, 1992). In Foucault's eyes, the soul is the prison of the body. The soul (or mind) is what is produced when the body is worked upon and disciplined in specific ways. This can be called a physicalistic aspect of Foucault's argument.

Foucault's point is that the self-monitoring subject was constituted in modernity through a number of institutional practices: in schools, factories, prisons, hospitals, courts of law, and in the whole system of treatment of individuals at large. The soul, the inner psychological world, became the *panopticon* that we all carry within ourselves (Coles, 1992), which is an important aspect of the current psychological social imaginary. Psychology became the science of the inner world, and therapies and other similar power techniques became normalizing, subjectivity constituting social technologies, especially when Western men and women became what Foucault called confessing animals, constituted as subjects through confessional technologies (Foucault, 1980). Thus, we *are* not simply subjects. We are *made* subjects through processes of subjectification.

In the final books and articles written by Foucault before his death in 1984, his interests shifted from external power techniques to what he called technologies for individual domination (Foucault, 1988a). He became interested in "the history of how an individual acts upon himself, in the technology of self" (p. 19). Technologies of the self are tools with which an individual acts on herself to create, recreate and cultivate herself as a subject (e.g., the Stoic practice of letter writing, Augustine's confessions, examinations, asceticism, and interpretations of dreams).

Contra historical materialists, Foucault claimed that technologies of the self are relatively independent of their socio-economic and political conditions. For that reason, humans are much freer than they think (Foucault, 1988c). The consequence of the fact that the self is not given us to discover in any universal way, is that we must create the self as a work of art (Foucault, 1984). Technologies of the self thus need not be oppressive. They can be linked with ethics as practices of freedom. By "ethics," Foucault meant not the abstract philosophical discipline, but a practice, viz. the practice of subjective self-formation. Ethics is the practical relationship a self ought to have to itself, and, in this sense, it can be studied independently of moral codes, and also independently of scientific truth. There is no connection at all, according to Foucault, between ethical problems and scientific knowledge. We can just as little demand that ethics should be true scientifically as we can demand that a work of art should be based on scientific criteria.

That is why we must create ourselves as a work of art, according to Foucault. There is no deep, inner truth about human nature, or about individual subjectivity to be unearthed by scientific psychology, so the goal of practicing technologies of the self should not be to decipher such an illusory truth. If this happens, then technologies of the self become oppressive. Then they come to belong in the hands of scientific experts – "masters of truth" – and then such technologies degenerate into techniques of domination. That is why Foucault wants to replace the classical Western ideal *know thyself* with what is for him a more primary demand: *Care of the self* (Foucault, 1988a, 1990).

The Techne of Psychology

The central point in Foucault's view of the human subject is that individual human beings are constituted as subjects (subjects to be known and controlled, and subjects who know and control) by practical technologies, infused in power relations. This is Foucault's version of how psychology and other human sciences are able to constitute human subjects and make human kinds loop: By providing society with technologies for subjectivity constitution. Nikolas Rose has continued Foucault's project by analyzing, inspired by Foucault's (1994a) notion of governmentality, how "governmentality or simply government have become 'psychologized.' The exercise of modern forms of political power has become intrinsically linked to a knowledge of human subjectivity" (Rose, 1996b:117). Government refers to the strategies for "the conduct of conduct" (p. 116) that have proliferated over the past two centuries, significantly supported by the psy-knowledges, the disciplinary regime of psychology (Rose, 1985, 1996a, 1999a), and more recently by the bio-sciences (Rose, 2007). Foucault and Rose intend the notion of governmentality to connect broader forms of power and political governmental control with finer micro-processes of power (Hook, 2003), for example in psychological practices such as therapy or testing.

Rose's analyses deal with the *techne* of psychology (Rose, 1996a, 1996b). In Aristotle's classical usage, *techne* is the kind of knowledge needed to produce things, and the *techne* of psychology is thus the kind of knowledge needed to produce subjectivities. Rose wants to consider psychology "not merely as a body of thought but as a certain form of life, a mode of practicing or acting upon the world" (Rose, 1996b:116). Living a life, Rose argues, has become a psychological and even a therapeutic business. When living according to the psychological social imaginary, we are obliged to make our lives meaningful through the search for happiness and self-realization in our individual biographies, and psychological techniques have become imperative in this regard. The self has become a commodity or a raw material to be worked upon, and psychological expertise is now central in devising ways of working upon the self. The goal is freedom, the modern liberal self is "obliged to be free" (Rose, 1996a:100), and psychological expertise and power have become "ethicalized" (p. 92), linked to authority concerning the good, meaningful life.

According to Rose, the key to understand psychology and its history lies in understanding how it has served to assemble human subjects through its technologies. Rose develops what he calls "an analytic of human technologies" (Rose, 1996a:27) where looping effects, the ways psychology affects its subject matter, are conceived as a technological affair. On this point, Rose is critical of what he sees as social constructionists' emphasis on meaning, narrative and linguistic discourse[5]:

> Subjectification is not to be understood by locating it in a universe of meaning or an interactional context of narratives, but in a complex of apparatuses, practices, machinations, and assemblages within which human being has been fabricated, and which presuppose and enjoin particular relations with ourselves (Rose, 1996a:10).

"Technology" refers to "any assembly structured by a practical rationality governed by a more or less conscious goal. Human technologies are hybrid assemblages of knowledges, instruments, persons, systems of judgment, buildings and spaces" (Rose, 1996a:26). The normative aim in Rose and Foucault is not to liberate humans from the thralldom of technology, for, in their eyes, subjectivity is necessarily and always a technological assemblage. Rather, it is to question the "more or less conscious goal" that informs the technologies through which humans are currently assembled as psychological subjects (particularly the liberalist goal of governing autonomous selves), and the ways our identities and moralities consequently have been bound to an inner psychological realm. However, I believe that "an analytic of human technologies" and the view of ethics and subjectivity developed by Foucault, has no answer to give us concerning an ethic of existence more adequate than the one currently informing the psy-sciences. For we are given no resources for evaluating whether psychological technologies affect us in good or in bad ways other than a turn to the individual's subjective self-formation. As we shall see in the following, the Foucauldian analytic of human technologies cannot account for the role values and meanings play in social practices, however rich it may otherwise be in its critical potentials, and it cannot give us grounds for a moral assessment of psychology's products.

---

[5] Foucault himself argued that subjectivity "is constituted in real practices – historically analyzable practices. There is a technology of the constitution of the self which cuts across symbolic systems while using them" (Foucault, 1984:369). One is not made a subject through language and discourse alone, as some constructionists, but also deconstructivists tend to claim. Foucault dismissed Jacques Derrida's textualization of practices and the Derridian claim that there is nothing outside the text. In a response to Derrida, he calls such textualization a "little pedagogy" that gives its practitioners "limitless sovereignty" by allowing them "to restate the text indefinitely" (quoted from Flyvbjerg, 2001:115).

Hermeneutic Processes of Self-Interpretation

I will now contrast Foucault and Rose's "technological" approach to human subjectivity and the looping effect of human kinds with the alternative interpretive-pragmatic approach, which understands humans as self-interpreting beings. In introducing the notion of looping effects of human kinds, Hacking originally did not want to enter the *Verstehen-Erklären* debate in psychology and social science: "The *Verstehen* dispute has partly to do with methodology, a subject that I abhor [...] I believe there are some deep insights on the *Verstehen* side of the argument, but here they are irrelevant" (Hacking, 1995b:364). I believe, however, that Hacking's view of things could become (even) more interesting to psychologists, if it were stated with reference to hermeneutic philosophy. As he says of the looping effect of human kinds: "people are aware of what they are called, adapt accordingly, and so change, leading to revisions in facts and then knowledge about them" (Hacking, 1995b:388). This "being aware," which humans are capable of, is exactly what characterizes our existence as *Dasein*, according to Heidegger's (1927) ontological hermeneutics[6] (cf. Brinkmann, 2004b; Stenner, 1998). This has also been argued by Martin and Sugarman (2001), who use Hacking's discussion of human kinds as a springboard for developing a hermeneutic approach in psychology. They claim that human kinds demand a hermeneutic understanding, because, although human kinds are definitely real, "it is a reality in which they themselves are deeply involved" (p. 194). The idea of humans "being involved" in their own reality is the focus of ontological hermeneutics as it has been developed by Heidegger, Gadamer, and Taylor among others. While Foucault was primarily engaged in the problematization of practices, and would leave the individual with the task of creating herself as a work of art, hermeneutic writers tend to deemphasize individual choice and claim that the values of practices cannot be accounted for as something subjective. Instead they typically point to communities and traditions as the source of meanings and values.

Ontological Hermeneutics

Originally, hermeneutics was developed as a methodology for interpreting texts, notably biblical texts (see Chap. 9 in Richardson et al., 1999, for an illuminating history of hermeneutics). With Wilhelm Dilthey in the late nineteenth century, hermeneutics was extended to human life itself, conceived as an ongoing process of interpretation. Heidegger's '*Being and Time*' (Heidegger, 1927) is often cited as the work that inaugurated the shift to ontological hermeneutics proper. The question

[6] Heidegger's hermeneutics is *ontological* because it is not concerned with the epistemological question of finding a correct method of interpretation or making sure that we understand others, but rather inquires into "the mode of being of the entity who understands" (Richardson, Fowers, & Guignon, 1999:207).

of methodological hermeneutics had been: How can we correctly understand the meanings of texts? Epistemological hermeneutics had asked: How can we understand our lives and other people? But ontological hermeneutics – or "fundamental ontology" as Heidegger also called it (p. 34) – prioritizes the question: "What is the mode of being of the entity who understands?" (Richardson et al., 1999:207). *Being and Time* aims to answer this question and can thus be said to be an interpretation of the act of interpretation (p. 208).

Heidegger's name for the entity that understands is *Dasein*, and the being of *Dasein* is unlike the being of other entities in the universe. Physical entities such as molecules, tables, and chairs are things that have categorical ontological characteristics, whereas human beings (*Dasein*), are *histories* or *events* and have existentials as their ontological characteristics (Polkinghorne, 2004:73f). *Dasein* primarily exists as involved in a world of meanings, relations and purposes, and only derivatively in a world of physical objects. *Dasein* lives in a normative space of reasons (to speak the language of Sellars and McDowell). In our everyday lives, we live absorbed in pre-established structures of significance, or what Heidegger called *equipment*, which serve as a background that enables specific things to show up as immediately meaningful and value-laden, given our participation in different social practices. We do not experience meanings and values as something we subjectively project unto the world, for the world in which we live meets us as always already imbued with meaning and value. Only when our everyday, unreflective being-in-the-world breaks down, when our practices of coping with the equipment somehow become disturbed, do entities appear with "objective" characteristics distinguishable from human subjects.

For Heidegger and later hermeneuticists like Gadamer and Taylor, understanding is not something we occasionally do, for example, by following certain procedures or rules. Rather, understanding is the very condition of being human (Schwandt, 2000:194). We always see things *as* something, human behavior *as* meaningful acts, and letters in a book *as* conveying some meaningful narrative. This, however, should not be understood as implying that we normally make some sort of mental act in interpreting the world. "Interpretation" here is not like the mental act of interpreting a poem, for example. It is not an explicit, reflective process, but rather something based on skilled, everyday modes of comportment (Polkinghorne, 2000). In the ontological hermeneutical perspective, interpretation is not so much something we *consciously* do, as something that is pre-reflectively lived and depends on a tacit meaning-giving background. Interpretation depends on certain *prejudices*, as Gadamer famously argued, without which no understanding would be possible (Gadamer, 1960). Knowledge of what others are doing, and also of what my own activities mean, "always depend upon some background or context of other meanings, beliefs, values, practices, and so forth" (Schwandt, 2000:201). There are no fundamental "givens," for all understanding depends on a larger horizon of non-thematized meanings. So instead of "givens," we ought to talk about "takens," as Dewey argued, for qualities are "*selected* from this total original subject-matter which gives the impetus to knowing; they are discriminated for a purpose" (Dewey, 1929:178).

Self-Interpretation

According to Taylor's reading of Heidegger's fundamental ontology, we are self-interpreting animals (Taylor, 1985b). This means that humans *are* what they interpret themselves to be. This does not imply that any individual person can make himself Napoleon by trying to interpret himself that way, but rather that human communities of interpreters and their historical traditions constitute the meanings of our self-interpretations. Gadamer has said:

> In fact history does not belong to us; we belong to it. Long before we understand ourselves through the process of self-examination, we understand ourselves in a self-evident way in the family, society, and state in which we live. The focus of subjectivity is a distorting mirror. The self-awareness of the individual life is only a flickering in the closed circuits of historical life. *That is why the prejudices of the individual, far more than his judgments, constitute the historical reality of his being* (Gadamer, 1960:276–277).

As Hacking pointed out, there must be access to specific fields of descriptions if we are to be able to act intentionally in specific ways. And these fields of description that constitute our self-interpretations only exist in historically evolved traditions. Gadamer argues that this makes the condition of human and social science quite different from the one we find in the natural sciences "where research penetrates more and more deeply into nature" (Gadamer, 1960:284). In the human and social sciences, there can be no "object in itself" to be known (no natural kinds) (p. 285), for interpretation is an ongoing and open-ended process that continuously reconstitutes its object. The interpretations of social life offered by researchers in psychology and other human and social sciences are an important addition to the repertoire of human self-interpretation, and powerful fields of description offered by human science, such as psychoanalysis, can even affect the way whole cultures interpret themselves. This means that "social theories do not simply mirror a reality independent of them; they define and form that reality and therefore can transform it by leading agents to articulate their practices in different ways" (Richardson et al., 1999:227). On the face of it, this seems not to contradict the Foucauldian notion of how different practical technologies inscribe themselves on embodied subjects, thus disciplining and subjectifying them in specific ways. So what is the difference?

Values and Practices

The difference is that ontological hermeneuticists such as Taylor believe that we can to some extent articulate the background practices that constitute our self-interpretations, and, in doing so, get in touch with the values and goods that define them as practices. This is in contrast to Foucault, who portrayed background practices (without using this term) as a field of power, mediated by technologies. Foucault did not want to articulate the meanings and values of practices, but rather to problematize them (Stern, 2000). His account of the rise of disciplinary power, for example, does away with all references to human values and intentions, and portrays disciplinary practices as just growing in modernity. According to Taylor, "Foucault was profoundly uninterested in any conception of the good order. There is none such. There is, in particular, no solution to the social problem of a good order. The good is all in the register of individual autonomy" (Taylor, 2000:133).[7] But we can, Taylor contends, articulate the reasons why practices grow, and thereby refer to the intentions and values that people have wanted to realize. There are no doubt good reasons to criticize these intentions and values, but the point is that if we want to understand these practices, we have to refer to the values that guide them. Social practices presuppose both materiality *and* value-laden human self-interpretations (Taylor, 2004:31).

In the hermeneutic perspective, the background that constitutes our subjectivities is not just "technological" but also meaningful, infused with values and affected by human intentions. Practices involve "oughtness" and social imaginaries (Taylor, 2004), and are only possible if they "make sense" in some way, and this they do only by incorporating values, self-conceptions and modes of understanding. Taylor argues that we can sometimes change our background practices by articulating the values that inform them. For example, the modern atomistic view of society is indeed, in Taylor's view, guided by a genuine human value, viz. freedom. In articulating practices of freedom in modern societies (or practices of subjectification, as Foucault saw them) we must refer to the value of freedom if we are to understand them. The way to criticize them is not to do away with all references to normative values, but rather to see if they realize the values on which they are purportedly built. And then we might find that a truly atomist society in fact needs "a maximum of bureaucratic surveillance and enforcement to function" (Taylor, 1985c:110). This then defeats rather than realizes the value of freedom. Thus, ontological hermeneuticists reject the atomist view of society by way of immanent critique (background articulation), rather than Foucauldian problematization.

I share Taylor's worry that Foucault's problematization of practices collapses into moral subjectivism; the view that all moral orders are equally arbitrary (Taylor, 1989:99). Taylor sees Foucault's late turn towards an "esthetics of existence" and "the self as a work of art" as a way to restate the idea that the good is above all about individual autonomy (Taylor, 2000:133). According to Taylor, Foucault offers "charters for subjectivism and the celebration of our own creative power at the cost of occluding what is spiritually arresting in [...] contemporary culture" (Taylor, 1989:490). With Foucault and also Rose we are offered no way of assessing what is valuable and what is dangerous in the ways psychology works to constitute its own subject matter, other than the rather empty one of letting the individuals decide what should contribute to creating their selves as works of art (see also

[7] It should be borne in mind that there are other contradictory interpretations of Foucault, for example Olssen's (2002), who sees Foucault as advocating a normative theory, which Olssen labels "thin communitarianism." I find Olssen's interpretation unconvincing, and he also admits that Foucault's notion of community is "untheorized" (p. 490).

Taylor, 1984). There is no recognition of the values on which practices necessarily operate, and this is why the pure "analytic of human technologies" is inadequate from the interpretive-pragmatic viewpoint.

If Foucault's esthetization of existence can be characterized as subjectivist and morally problematic, what resources for evaluating practices, including the ones informed by psychology's operations, are then available from the interpretive-pragmatic viewpoint? Traditionally, three non-subjectivist answers have been given. First, the classical Aristotelian answer that points to an essential, ahistorical human nature that can be either nourished or thwarted by the theories we make about it. Second, the rationalist or the Kantian answer that points to rational and universal procedures for securing justice. A significant theory here is Jürgen Habermas's discourse ethics (e.g., Habermas, 1993). Although Aristotle's conception of practical wisdom has inspired many hermeneutic and pragmatist philosophers, including Heidegger, Gadamer, and Dewey, and even though Habermas can be counted among hermeneutic and perhaps also pragmatist theorists, most contemporary hermeneuticists and pragmatists argue that neither human nature (at least in the essentialist, Aristotelian sense) nor universal procedures of rational discourse can give us the moral criteria required to evaluate the looping effects of human kinds. Human nature (as "first nature") is too thin to give us guidance, and the Habermasian procedures of discourse ethics are often perceived as too rigid, too objectifying, and too ethnocentric. (I discuss these problems further in part II of this book.)

The third interpretive-pragmatic answer points to the historical development of practices as the source of moral guidance.[8] We can view Taylor's entire philosophy as a way to articulate this third alternative. For Taylor, good interpretations in social science are those that enrich our practices by enabling us to lead better lives. According to which criteria? As Schwandt says, "For Taylor, what counts as better interpretation is understood as the justified movement from one interpretation to another" (Schwandt, 2000:202). Taylor would reject Habermas's view that we can begin from nowhere every time we evaluate some interpretation or judgment. Instead, practical reasoning is always, Taylor argues, comparative or reasoning in transition (Taylor, 1995a). We can demonstrate that a given position is superior to a rival, but we cannot show it to be a true *simpliciter* (Brinkmann, 2004b:71).

When we articulate the background that shapes our ideas about ourselves, we are engaged in an exercise in practical reason, but one wholly unlike modernity's instrumental reason. Articulating the background "may open the way to detaching

[8] Teehan's reading of Dewey's pragmatism and Taylor's hermeneuticism is relevant here. He argues that both Dewey and Taylor find that there "is no ultimate, foundational basis for judging one position better than another, but that this does not leave the choice between positions arbitrary" (Teehan, 1996:88). Teehan even argues that Taylor (perhaps unknowingly) "is advocating a Deweyan theory of inquiry" (p. 88). Both Taylor and Dewey claim that understanding the world presupposes a background of tradition, prejudices and habits, both see social theory as a practice, both argue that understanding ultimately involves application, and both find that the human power of reasoning is a sort of situated reasoning in transition.

ourselves from or altering part of what has constituted it – may, indeed, make such alteration irresistible; but only through our unquestioning reliance on the rest [of the background; my addition]" (Taylor, 1995c:12). We inevitably criticize our self-interpretations, including those that modern psychology has produced from within certain other self-interpretations that exist as parts of historically evolved traditions. When Foucault recommends that we create ourselves as a work of art, he forgets that for something to count as a work of art, a tradition of art appreciation is required with certain normative standards. We simply cannot escape normativity. Obligations towards our society and its social relations precede the individual's own freedom and self-formation since such relations are what make situated individual freedom possible in the first place (Taylor, 1985a:209). In other words, duties are prior to rights (Harré & Robinson, 1995). In addition to pointing to the necessary reflection on our cultural embeddedness, Taylor has offered a number of other ways of establishing ways of evaluating interpretations, for example when we try to understand and evaluate the self-interpretations of another society. In this case, we should work out "a language of perspicuous contrast" in which we can "formulate both their way of life and ours as alternative possibilities in relation to some human constants at work in both" (Taylor, 1981:205). This is close to Gadamer's notion of a fusion of horizons, and Taylor believes that this will permit us to rightfully criticize the practices of other cultures, as well as those of our own.[9]

From his interpretive viewpoint, Gadamer concluded that the human sciences are inescapably moral sciences:

> The human sciences stand closer to moral knowledge than to [...] "theoretical" knowledge. They are "moral sciences." Their object is man and what he knows of himself. But he knows himself as an acting being, and this kind of knowledge of himself does not seek to establish what is. An active being, rather is concerned with what is not always the same but can also be different. In it he can discover the point at which he has to act. The purpose of his knowledge is to govern his *action* (Gadamer, 1960:314).

Gadamer views knowledge in the social sciences as *practical* knowledge, and he emphasizes the dimension of *application* in all human understanding. Hermeneutics is therefore *practical philosophy*, and Gadamer points to Aristotle's conception of practical wisdom – *phronesis* – as the model for hermeneutic understanding. Understanding is always bound up with application, and, in *Truth and Method*, Gadamer develops this view from legal hermeneutics, where understanding the law necessarily involves applying it to concrete cases. One does not really understand the law if one is unable to apply it. New applications again develop the practice of law creatively, just as future instances of legal judgment must refer back to earlier

[9]In "Explanation and Practical Reason" (Taylor, 1995a), a number of other ways of evaluating competing interpretations, as a kind of reasoning in transitions, are delineated. When interpretations or theories are in disagreement, the superior theory will be the one that makes better sense of inner difficulties than its rival, or presents a development which cannot be explained on the terms of the rival, or shows that it itself has come about through a move that can be intrinsically described as error-reducing compared to its rival (p. 54).

instances, that is, to the *tradition* of law. And so it is with human science in general, Gadamer argues. In understanding what humans do, we apply certain historically evolved descriptions to their activity, and, as acting and self-interpreting beings, humans can then employ social researchers' understandings as self-understandings.

Taylor has developed this view further in his thesis that social theory is a kind of practice (Taylor, 1985c). It is a practice that serves to interpret and articulate the meanings of human activity, but these articulations may again enter the actors' self-understandings, thereby changing the realities they are concerned with. Taylor's point is that validity in the social sciences cannot mean mirroring some independent objects researched, for the objects of human and social science are *not* independent of human understanding (but exactly *constituted* by such understanding). Thus, validity in the social sciences means *improving* the practices under consideration, and this is a moral issue. New interpretations can alter the self-understandings of those they describe, and social theories can thus be tested by examining the quality of the practices they inform and encourage. This is a pragmatic element in Taylor's hermeneuticism.

The interpretive approach of Gadamer and Taylor differs from the social constructionist in their argument that we can rationally judge and compare the quality of different theories and practices. This is not, however, from some standpoint wholly external to human history and meanings, but from *inside* lived practices, which was also argued earlier (and is further spelled out in Brinkmann, 2004b:70–72). Social science as practical reason is, as Taylor says, *reasoning in transition* (Taylor, 1995a). We can make the transition, for example, from paternalistic theories about women to theories stressing the worth and equality of women and men, while representing this as a clear *gain* in understanding, whereas it seems impossible to go in the other direction. Such a comparison is an exercise of practical reason which, Taylor and Gadamer think, cannot be put on theoretical formula, but depend on the character of the interpreter. As Taylor put it in an early article, referring to the social sciences:

> These sciences cannot be "*wertfrei*"; they are moral sciences in a more radical sense than the eighteenth century understood. Finally, their successful prosecution requires a high degree of self-knowledge, a freedom from illusion, in the sense of error which is rooted and expressed in one's way of life; for our incapacity to understand is rooted in our own self-definitions, hence in what we are. To say this is not to say anything new: Aristotle makes a similar point in Book I of the *Ethics*. But it is still radically shocking and unassimilable to the mainstream of modern science (Taylor, 1971:51).

Conclusions

I have attempted to answer two questions in this chapter: First, what does it tell us about the nature of the subject matter of psychology, if the discipline is involved in "making up people?" Second, how should we understand the processes through which such "making up" occur? In response to the first question it was argued that it tells us that psychology's objects are human kinds, which interact with their classifications and are able to exert influence on themselves. Unlike natural kinds, human kinds demand social discursive contexts and relevant fields of description in order to exist. In short, they demand a space of reasons and an evaluative background. In response to the second question, I investigated two different ways of answering it, the Foucauldian and the interpretive-pragmatic (in this case I primarily drew on hermeneutic philosophy). Below are listed the main features of their answers, and how they each frame the workings of psychology:

Foucauldian perspective: Psychology has come to constitute our subjectivities through a complex historical power play, which has tied us to the project of our own identities in the hands of psychological "masters of truth." Rather than articulating what we are, we should problematize and "refuse what we are" (Foucault, 1994b:336), after having realized that "the self is nothing else than the historical correlation of the technology built in our history. Maybe the problem is to change those technologies" (Foucault, 1993:222f). We should face the fact that psychology is a *techne* that produces human subjects, and, in trying to change the technologies that shape us, we should embrace an "esthetics of existence" and create ourselves as a work of art.

Interpretive-pragmatic perspective: Psychology has come to shape our background understandings (of what is meaningful, of what is worthwhile, of what is valuable, of what we are) and thereby come to influence our self-interpretations and the social imaginary. Today we therefore pre-reflectively relate to ourselves and others in psychologically informed ways. We imagine life in psychological terms. The space of possible intentions and actions has been significantly affected by the new fields of interpretation opened up by psychology. By articulating the ways psychology has informed our background understandings (thereby formed us as agents), we are engaged in a process of practical reasoning, viz. in investigating if the values that guide us are worth being guided by.

From the hermeneutic critique of Foucault's "physicalism" – that it does not take into account that human actions and practices are guided by normative orders and real notions of value – I have argued that although the Foucauldian analytic is rich in its critical potentials, it is inadequate as the basis for a constructive approach to develop psychology as a moral science. In this light, Foucault's approach is morally subverting, and, in a paradoxical way, comes to support some of those aspects of modern social science that it sets out to criticize: viz. its commitment to a kind of value-neutrality. The Foucauldian problematization of social practices leads to the notion, at least as perceived by Taylor, that values are nothing more than subjective projections *ad modum* Nietzsche. Foucault was deeply suspicious of any perspective that would posit the human being as a detached, meaning-giving subject (such as Husserlian phenomenology). However, as Dreyfus and Rabinow (1982) spell out, ontological hermeneutics (along with pragmatism, I would add) "gives up the phenomenologists' attempt to understand man as a meaning-giving subject, but attempts to preserve meaning by locating it in the social practices and literary texts which man produces" (p. xv). This is in contrast to the Foucauldian strategy of problematization that does away with meaning and value.

But what does this have to do with morality? Let us see what Hacking has to say about this issue: In response to a question about whether our being aware of looping effects can provide a condition for making the quest a moral one for the betterment of humankind, Hacking answered, "Yes, this is the expression of a noble hope," and he added: "the looping effect is not necessarily a bad thing. Liberation movements of which many of us approve are part of the game" (Hacking, 1995b:394). Hacking is clear that human kinds are morally laden: "Human kinds are kinds that people may want to be or not to be, not in order to attain some end but because the human kinds have intrinsic moral value" (p. 367). Thus *the looping effect in psychology is a moral process*. By turning to the hermeneutic (and the interpretive-pragmatic) perspective, I believe psychologists can become better equipped to understand the moral values of their human kinds, perhaps after having appropriated the valid critical perspectives of Foucault and his followers.

Part II
An Inescapable Morality

In part I, I argued that many of us live our lives according to a psychological, social imaginary. We tend to imagine social life and its demands in psychological terms, and I argued that this has drastically foreshortened our conceptions of the moral domain. There is little room for value in a world that has been influenced by psychological modes of understanding. We often see morality as a psychological phenomenon, as connected to the inner workings of the mind, rather than as demands that arise "externally" from our dealings with each other and the world. We often use psychological techniques to reach our goals, without questioning whether our goals are worth reaching. We also saw that psychology can be considered as a significant cultural agent that continuously "makes up people," which ought to alert us to the moral dimensions of psychology.

In this second part, I take more constructive steps and begin to outline a way to think about psychology's subject matter that respects its moral nature. In addition to examining how psychology has affected our moral space, I now focus on the moral space itself with the hope of contributing to the development of a psychology that respects the inescapable moral normativity of human life. While part I was constructionist and historicist concerning the workings of psychology, I argue in the second part that we need a firmer realist stance concerning moral phenomena, some of which are non-constructed and foundational for more historically transient psychological objects. First, I tackle the thorny issue of the naturalistic fallacy, i.e., the objection that it is illegitimate to infer evaluative statements from descriptive ones. Then, in the following chapter, I defend a version of moral realism, the view that moral phenomena, values, and properties are real. I argue that moral phenomena are not figments of the imagination, subjective projections unto a value-neutral world, or something humans make up by an act of the will. Moral phenomena and moral properties are discovered rather than invented, and they are a precondition for the existence of a number of central psychological phenomena. In the subsequent chapter, I argue that morality is best thought of as embedded in social practices, and in the concluding chapter, I summarize my arguments and try to challenge them with a number of objections that have not been raised earlier in the book.

Chapter 5
Facts, Values, and the Naturalistic Fallacy in Psychology

Introduction

The relationship between facts and values is tremendously intricate in psychology, as in the other human and social sciences, which take as their object of interest human beings as acting persons. Often, two premises are taken for granted in scientific psychology: First, that only statements of fact can be objectively true, while statements of value can be nothing but expressions of subjective preferences. Second, that there is an unbridgeable logical gap between factual and evaluative statements so that no descriptive statement can entail an evaluative statement (without the addition of some evaluative premise).

It was David Hume, the grandfather of empiristic and "value neutral" psychology, who famously observed that an "ought" cannot be logically derived from an "is." In his *Treatise* from 1739 he notes that in all the systems of morality he has seen, the author shifts suddenly from "is" and "is not" to "ought" and "ought not" (e.g., from "God *is* our Creator" to "we *ought* to obey him"), without explaining or justifying this "new relation" (Hume, 1978:469–470). At the beginning of the twentieth century, moral philosopher G.E. Moore introduced the term "the naturalistic fallacy" for the attempt to derive evaluations from natural matters of fact. Unlike Hume, Moore was no subjectivist regarding moral judgments. He argued that the reference of moral terms could not be subjective psychological states such as emotions or sentiments (Robinson, 2002:28). Moore believed that goodness must be a simple, non-natural property that cannot be defined in terms of other (natural) properties (see the illuminating account in Casebeer, 2003). Moore's argument is known as the open-question argument: If goodness were definable in terms of some other property, as the so-called naturalists believe, then we would consider it an illegitimate question to ask whether this property really was good. It would be like asking whether cats are really cats. But, as Moore observed, it is in fact meaningful to pose such questions. If a utilitarian philosopher argues that preference-satisfaction should be the basis of an ethical system, for example, it still makes sense to ask, "but is preference-satisfaction really and always good?" If "good" simply meant "preference-satisfaction," then the sentence "what satisfies our preferences is good" would not inform us of anything, or give us additional

S. Brinkmann, *Psychology as a Moral Science: Perspectives on Normativity*,
DOI 10.1007/978-1-4419-7067-1_5, © Springer Science+Business Media, LLC 2011

reasons to endorse preference-satisfaction. Moore himself argued that "good" simply cannot be defined empirically, but instead is apprehended through intuition.

It seems possible to treat Moore's open-question argument as a species of the more general thesis of the naturalistic fallacy (Casebeer, 2003), and, in this chapter, I shall focus on the latter and its relation to psychology. I shall attempt to question the thesis that one cannot derive values from facts by drawing on four different arguments that are all relevant to psychology. These are not unrelated, although they have been developed within somewhat different sets of literature. By presenting these arguments, I hope to provide a map of different ways of "re-moralizing" or "de-psychologizing" psychology. I believe that an adequate perspective on the relationship between facts and values in psychology can help move the discipline toward a science of living and acting human beings that is true to the value issues that seem to be an indispensable part of our reality (Robinson, 2002).

The four arguments discussed below lead me to propose that (1) we have a capacity for perceiving "oughtness" as part of the world; (2) given that rules and norms order social life, there seems to be an inherent "factual normativity" in our lives; (3) human functioning in general cannot be understood in value neutral terms; and (4) value terms cannot be exorcised from our vocabulary, at least not from the part of our vocabulary that designates action, thinking, and feeling. Together, the arguments show that much of what psychologists study, our perception of the world, our ordered social reality, our functions as human beings, and our discursive practices, could not be what they in fact are, if there had been an unbridgeable gap between facts and values, as presupposed by the thesis of the naturalistic fallacy. For each of the arguments, I discuss counter-arguments from positions in favor of the fact-value dichotomy and the naturalistic fallacy, but I hope to demonstrate that these are ultimately unsuccessful. While the naturalistic fallacy has been discussed before in psychology (e.g., Kohlberg, 1971), I hope in this chapter to also move beyond that discussion by proposing that a central question for those psychologists, who believe that values are a real and central part of human life, is whether the normativity of values should be grounded in human functions that have evolved in the course of our natural history (Casebeer, 2003), or in the more variable social practices that exist in human cultures (MacIntyre, 1985a). I argue in this chapter and the following ones that social practices are the main source of normativity – moral and otherwise – in our lives, and I propose that the four arguments invoked to illustrate the problems of the so-called naturalistic fallacy can be explained by understanding perception, social rules, human functions, and thick description as aspects of the fundamental field of practices that makes up our living social and psychological reality.

The Fact-Value Distinction in Psychology

Before presenting the arguments against the thesis of the naturalistic fallacy, however, it is relevant to look briefly at the history of the fact-value distinction in psychology. Nineteenth century philosophical psychologists like John Stuart Mill still prided themselves on belonging to "the moral sciences," and early psychologists such as Wilhelm Wundt and Eduard Spranger did not hesitate to investigate values. Wundt even wrote an *Ethik*, and he argued that there is no psychological structure that is free of values (Giorgi, 2006:7). Spranger, who was a student of Wilhelm Dilthey, investigated people's personal values, and his ideal types of different values inspired the Allport–Vernon–Lindzey Study of Values test in the 1950s. Many other prominent psychologists were deeply interested in moral issues, among them William James, Sigmund Freud, B.F. Skinner, and, perhaps surprisingly, Hermann Ebbinghaus (Giorgi, 1992). In particular, the school of Gestalt psychology placed values at the heart of human psychological life (to be discussed below). In addition, J.J. Gibson's ecological approach aimed to develop an ontology for psychological phenomena that placed values in the world. Gibson explained that he had "been moving toward a psychology of values instead of a psychology of stimulus" (cited in Hodges & Baron, 1992:263), and, in his ecological approach, the "perceiving of an affordance is not a process of perceiving a value-free physical object [...] it is a process of perceiving a value-rich ecological object [...] Physics may be value-free, but ecology is not" (Gibson, 1986:140).

Furthermore, there have been the well-known empirical investigations of moral judgment and reasoning originating from Piaget's *The Moral Judgment of the Child*, Kohlberg's studies of moral development, and Gilligan's feminist critique. Few, however, have discussed the basic issues concerning the naturalistic fallacy, the ontology of values and their relationship to the subject matter of psychology. Kohlberg is an exception with his paper "From is to ought: How to commit the naturalistic fallacy and get away with it in the study of moral development" (Kohlberg, 1971). In the paper, Kohlberg asserts that he does *not* commit the naturalistic fallacy in the sense of deriving moral judgments from psychological judgments (e.g., pleasure–pain statements), or in the sense of finding the source of morality in human biological nature, as many of today's evolutionary psychologists attempt (e.g., Hauser, 2006). Instead, as a Kantian cognitivist, Kohlberg merely commits the naturalistic fallacy in the sense of asserting that "any conception of what moral judgment ought to be must rest on an adequate conception of what it is" (Kohlberg, 1971:222). Thus, he still respects a fundamental fact-value dichotomy and in a certain respect simply endorses Kant's famous dictum that ought implies can, which, in my view, is not really to commit the naturalistic fallacy.

Although these examples, and also other more recent ones (Lacey & Schwartz, 1996; Richardson, Fowers, & Guignon, 1999), demonstrate that psychologists have given the fact-value distinction serious and critical thought, it seems fair to conclude that the majority of psychologists qua scientists take for granted the two premises mentioned above: That only statements of fact can be objectively true, and that there is a logical gap between factual and evaluative statements. Howard Kendler, for example, has restated Hume's view by referring to the alleged "failure of *is* to logically generate *ought*" (Kendler, 1999:832). Science, says Kendler, "by itself, is incapable of converting empirical relationships into moral principles or social policies" (Kendler, 2002:491). Psychology is only competent to "estimate the consequences of different social policies" but cannot "identify the morally correct one" (p. 501). This seems to be a fairly standard view in psychology, informed in large measure by the thesis of the naturalistic fallacy.

All of this is not to say that psychologists do not investigate values, but when they do so, they often treat them as empirical facts, more or less like opinions or attitudes, which is to say that they often *psychologize* values by disregarding or reducing their normativity (Smith, 2006). In logic, as Husserl and Frege argued in the early twentieth century, psychologism is illegitimate, for it is not our mental operations that are the source of the validity of logical inference. Rather, logic is the normative standard in light of which we can evaluate the validity of psychological thought processes. Consequently, it seems to be impossible for psychologists to demonstrate empirically that people tend to commit certain errors (when engaged in logical thinking) without presupposing the basic normativity of logic: Namely that there are correct and incorrect ways of performing logical operations. In contrast, when the topic is values and moral thinking, psychologists often suspend their normative judgments, possibly because they do not think there are objective truths about values to be unearthed by science. To give just one illustration of what I mean by this, consider the following quote from Martin Seligman's *Authentic Happiness*, describing the task of his positive psychology:

> It is not the job of Positive Psychology to tell you that you should be optimistic, or kind or good-humored; it is rather to describe the consequences of these traits [...]. What you do with that information depends on your own values and goals (Seligman, 2002:129).

Values are here seen as something individual, like someone's personal goals, and psychology should not tell you what values to hold, but should, in order to avoid committing the naturalistic fallacy, restrict itself to factual descriptions of the consequences of having certain values. In the remainder of this chapter, I hope to give good reasons to believe that there is a sense in which we cannot do psychology without committing the naturalistic fallacy, for the factual subject matter of psychology is shot through with normative values.

The Phenomenological Argument

What I shall refer to as the phenomenological argument takes its lead from the omnipresent feature of our experience that we are perceptually confronted with demands to do certain things. I shall take Gestalt psychology as representative of the view that human beings – as creatures of interest to the science of psychology – are faced with objective value-based demands. Later, I present and criticize a currently influential counter argument to this approach, based on evolutionary theory, namely that perception of moral oughtness is a "functional fiction."

What does it mean to say that, phenomenologically, we are perceptually confronted with moral demands? What it means is very simple. In our everyday lives and in everyday conversations, for example, we are normally aware of what our words do to others, we notice when someone needs cheering up, when someone is being offensive, and when someone's feelings get hurt. Gestalt psychologist Wolfgang Köhler referred to these aspects as the "requiredness" of situations, and argued that we experience such requiredness directly (Köhler, 1944, 1959). We not only see "what *is*" in our environments, but also often "what *ought* to be." At the bottom of all human activities are values, Köhler (1959:35) argued, the conviction that some things "ought to be." It is part of our experience that we perceive these value qualities as immanent in objects and events, and, Köhler added, qualities belong where we find them. No scientific explanation can change a phenomenon or its location. Thus, it is not normally the case that we subjectively imbue things with value. If we correctly experience a value attribute in a thing, then it is the thing per se that has this attribute (Robinson, 2002:32). Moral ascriptions, according to the Gestaltists, are thus akin to naming colors or identifying musical harmony, and the result is, as Robinson says in agreement with Gestalt psychology:

> Deriving 'ought' from a complex pattern of events is no more puzzling, let alone fallacious, than to derive 'blue' from wavelengths, just in case the two kinds of events are comparably common in human experience owing to their being tied to features of the natural world (Robinson, 2002:37–38).

Köhler thought it crucial that value has a *demand* character, and does not depend upon acts of the self (Köhler, 1944:206). We experience ourselves being *moved* by values that are there independently of our subjective perspective. As Charles Taylor has put it: "We sense in the very experience of being moved by some higher good that we are moved by what is good in it rather than that it is valuable because of our reaction" (Taylor, 1989:74). According to the Gestaltists, values are therefore objective facts. Kurt Koffka distinguished between two meanings of the conceptual pair subjective–objective. On the one hand, it can refer to whether something belongs, or does not belong, to the self. Something is objective in this *phenomenal* sense if it does not belong to the self. But the conceptual pair can also mean dependent or not on organisms. Something is objective in this second *functional* sense if it does not depend on organisms (Koffka, 1940:192–194). For example, pain is subjective in both senses. Pain is necessarily someone's pain (and is thus phenomenally subjective), and it depends on the existence of an experiencing organism (and is thus also functionally subjective, according to Koffka). Colors and values, on the other hand, are functionally subjective like pains, but they are phenomenally objective. That is, there can be no colors or values without experiencing organisms, and yet they do not belong to the experiencing self. We cannot change them at will (below I shall take up the question of the reality of values in greater detail). According to Koffka, it is psychology's job among the sciences to deal with objects that are *functionally subjective* (e.g., colors, pains, values, thoughts, and emotions), and some of these are undoubtedly *phenomenally objective* (e.g., values and colors). If this is true, then there are phenomenological reasons why psychologists cannot accept the naturalistic fallacy or a strict fact-value dichotomy. For in characterizing the "requirednesses" of the world, we are engaged in factual description that is only possible given a normative background.

Needless to say, not everyone accepts this, and I shall now consider one currently influential objection to this conclusion, which comes from evolutionary theory. Of course, if the evolutionary perspective contains attempts to define the moral meaning of good as "survival" or "passing on genetic material," then it is itself a prime target for being accused of committing the naturalistic fallacy. This shall not concern me here, however, for at this point I am merely interested in its challenge to our moral phenomenology, stating that all this is *just* phenomenology: It may certainly *appear* to be the case, the objection runs, that we perceive moral oughtness as objective features of the world, but this is *mere* appearance rather than reality. To quote evolutionary ethicist Michael Ruse: "Even though morality may not be objective in the sense of referring to something 'out there', it is an important part of the experience of morality that we think it is" (Ruse, 1991:508). Our experiences of objective normative and moral properties are, from this point of view, *functional fictions* that serve evolutionary purposes. We are thus tricked into believing that values are "out there" rather than "in here." For, as Ruse claims

> if we recognized morality to be no more than an epiphenomenon of our biology [which Ruse thinks it really is], we would cease to believe in it and stop acting upon it. [...] It is important, therefore, that biology should not simply put moral beliefs in place but should also put in place a way of keeping them up. It must make us believe in them (Ruse, 1991:507–508).

Ruse's reduction of moral beliefs to functional fictions seems to be a powerful objection to the idea that moral ascriptions refer to real features of the world. This line of argument is akin to Mackie's (1977) case for the metaphysical queerness – and therefore non-existence – of values. Even though our moral discourse presumes that there are objective values to which we can refer, this presumption is simply an erroneous belief (possibly installed in us by evolutionary processes). However, as Nagel (1997) argues, such reductive strategies based on evolutionary theory are at risk of undermining themselves. The problem concerns their central claim that whatever reason we have to believe – morally and otherwise – is the result of our psychological apparatus as a response to evolutionary adaptation. However, if the reductionists want to remain consistent, this must also apply to the theory itself! On this evolutionary account, therefore, the only reason I could have to believe this account *itself* would be grounded in natural selection. Thus, if the evolutionary hypothesis itself depends on reason, and if reason is a product of natural selection, then the hypothesis is self-undermining. There must be something more than simply being a product of natural selection to human perception, thinking, and reasoning – moral and otherwise – if we are to trust these capacities. As Nagel says

> I can have no justification for trusting a reasoning capacity I have as a consequence of natural selection, unless I am justified in trusting it simply in itself – that is, believing what it *tells* me, in virtue of the *content* of the arguments it delivers (Nagel, 1997:136).

Currently influential attempts at undermining or reducing the phenomenological experience of oughtness are part of a more general attempt in philosophy to naturalize human powers. In epistemology, Quine (1969) and his followers stand as the most prominent exponents of a naturalized epistemology that seeks to replace epistemic normativity, concerned with justification, with a causal–nomological science of cognition. However, as Siegel (1996) points out – and I believe his argument applies to other attempts at naturalization as well, including the naturalization of values and morality:

> An argument for eliminating the normative must be an argument to the conclusion that it is in some sense more rational, or more reasonable, or better justified, to eliminate the normative from epistemology and to reconstrue the evidential relation as causal rather than as epistemic. But any such argument presupposes those very normative conceptions [...] that it seeks to eliminate (Siegel, 1996:7).

It thus seems that if we want to engage in rational discussion based on an exchange of reasons for beliefs and actions, we cannot leap out of the normative "space of reasons" in which we orient ourselves according to what is a reason for what (McDowell, 1994). Those who declare morality, and normativity more broadly, a functional fiction, approach the phenomenological "space of reasons" from sideways on, which disconnects them from the very discursive field where it makes sense to argue and convince others. In my view, there can be no purely descriptive science of values or morality from an outsider's perspective, for the very act of identifying something as moral is already a moral affair that has to be justified with good reasons (Davydova & Sharrock, 2003).

Although the discussion is far from settled with the phenomenological argument alone, I hope I have at least provided a first indication why a "scientific" elimination of the normative aspects of our moral perception, rendering it a "functional fiction," is much trickier than it may initially seem to its defenders.

The Constitutive Rule Argument

There are other ways to question the naturalistic fallacy, and I shall now refer to one of the most famous attempts to do so in modern philosophy. We shall see that if this argument is successful, it will also be relevant for psychology. In a now classic article, philosopher John Searle tried to demonstrate a counterexample to the thesis that one cannot derive evaluative statements from descriptive statements (Searle, 1967). From the statement "Jones uttered the words 'I hereby promise to pay you, Smith, five dollars'" (p. 102), which is quite clearly a descriptive "is" statement, we may validly conclude – via some deductive steps spelled out in great detail by Searle – that "Jones ought to pay Smith five dollars" (p. 102). Searle explains the validity of his deduction by invoking a distinction between "brute facts" (e.g., "a man has a bit of paper with green ink on it") and "institutional facts" (e.g., "a man has five dollars"). The latter is an institutional fact because it can only be a fact due to the institution of money. Likewise, a promise is only possible because of the institution – or the social practice – of promising. A promise would not make sense if only one person had ever promised anything and if all others were unaware of this institution (then, of course, it would not *be* an institution).

Be this as it may, a skeptic could still ask why it is not a completely subjective choice that one ought to live up to one's promises. Is the oughtness here not merely in the eye of the beholder? No, for what Searle calls "institutions" are possible only because of constitutive rules, which, like all kinds of rules, are public, at least in principle. As Wittgenstein (1953) demonstrated with his argument against the possibility of a private language, there is no such thing as following rules privately, that is, laying down rules that others are in principle debarred from knowing about. We can only follow rules if there is a *practice* of rule following in the relevant domain. Constitutive rules should be distinguished from those "regulative rules" that regulate activities whose existence is independent of the rules (Searle, 1967). An example of this could be the rules of polite table behavior that regulate the independent activity of eating. Constitutive rules, on the other hand, regulate *and* constitute the activity in question. The rules of football represent an obvious example. Without rules of the game, there would not *be* a game. Promising, money, marriage, etc. are institutions like football, according to Searle, that require a set of constitutive rules to exist and function. It is a constitutive rule that when someone promises something to someone, then he or she has undertaken an obligation – and it is a simple conceptual fact that one ought to do what one is under an obligation to do. This is how we explain the idea of obligation. Thus, even to recognize and identify something as a promise (in a factual or "descriptive" sense) involves granting that, all things being equal, it ought to be kept (in a normative sense) (p. 108). Thus, when we deal with institutional facts and constitutive rules, we may validly infer evaluative statements from descriptive ones.

At this point, we might relevantly ask what all this has to do with psychology. The answer is that if some psychological phenomena depend on the existence of "institutional facts," then these phenomena exist in a normative order from the outset that defies the naturalistic fallacy. As we saw in Part I, psychological phenomena (emotions, thoughts, actions) do require rules, norms, or conventions for their existence. Therefore, when doing psychology we should "refocus the search for *principles of order* from causality and causal mechanisms to conventions, customs, habits and practices […], and every practice is, in various ways, subject to normative appraisal" (Harré, 2004:6). This is so because practices (e.g., greeting someone, designing furniture) are not mere happenings, but activities that can be more or less correct, performed more or less well. In fact, we have reason to think that it is only possible to *identify* something as a psychological phenomenon on the background of some normative (i.e., rule-governed, normatively constrained) order. To reiterate the example discussed earlier, the reason why dread and anger are psychological phenomena (i.e., emotions), for example, but not indigestion or exhaustion – although all have behavioral manifestations as well as fairly distinctive experiential qualities – is that only the former belong to a moral order. We can be praised and blamed for being angry (there is intelligibly such things as justified and unjustified anger), whereas indigestion is outside our normative space of justification. So if the "patterns of episodes in which psychological phenomena are brought into being are expressions of locally valid norms, conventions and customs" (p. 4), then it seems that what Searle called constitutive rules are not just constitutive of a range of human institutions, but also of a range of psychological phenomena as such. These turn out to be incomprehensible if cleansed of their evaluative content.

I shall now discuss two challenges to this argument. The first says that all these appeals to institutions and norms do not matter at all, for psychology simply ought to get rid of any such references to intentions and norms, and, like physiology, become a purely causal science (I here skip the discussion whether physiology can be purely causal or whether it necessarily involves functional explanations). Again, Quine was an able exponent of this position, arguing that the social sciences qua sciences should strive to discover causal laws, and since the existence of causal laws are incompatible with the existence of intentionality and normativity, the latter aspects should simply be excluded from the sciences (a helpful discussion of this is provided by MacIntyre, 1985a:83). MacIntyre's rejoinder (like Harré's) states that if the dilemma is indeed as Quine presents it, then it is the search for causal laws, rather than the reference to intentions and norms, that ought to be excluded from psychology. For one does not explain the human world of actions, norms, and intentions by eliminating any reference to them; that would be to change the subject and miss the chance of understanding human life. We could of course try to prevent psychology from investigating its subject matter, but then we would simply need another discipline to do the job.

The second challenge I shall mention concerns the accusation of relativism that can be directed at the constitutive rule argument. If one is a moral relativist, then this accusation is not disturbing, but the problem is that those who question the naturalistic fallacy and are interested in "moral facts" and their psychological relevance cannot easily be relativists (for if all moral facts are relative, then there is a sense in which there are no such facts). Initially, the constitutive rule argument seems to lead to moral relativism, for if moral normativity is embedded in institutions, constituted by rules, and if such institutions differ across cultures and historical epochs, then normative morality apparently becomes completely relative to specific cultures and historical epochs. In an interesting footnote in his classic paper, Searle implicitly addressed this challenge when he considered whether it would be possible "to throw all institutions overboard" – in order perhaps to avoid having to derive an ought from an is – but this, Searle argues convincingly, would mean that we could not "engage in those forms of behavior we consider characteristically human" (Searle, 1967:113). If we are to have human institutions at all, we simply have to respect certain constitutive rules – and consequently, such respect becomes a universal moral value (Holiday, 1988). I return to this in the Chap. 6, when I unfold Holiday's Wittgensteinian defense of universal moral norms. I hope that I have here indicated why we should take seriously the idea that human psychology can only be understood by presupposing a non-relative evaluative background.

The Function Argument

What I shall here refer to as the function argument goes back to Aristotle, and relies on the premise that we only know what something essentially is, when we know its function (*energeia*). When we know the function of something, then we can investigate those excellences that allow that something to perform its function well.

"Excellence" is "*arete*" in Greek, which is sometimes translated as "virtue," but that is potentially misleading, since all things can display *arete*, not just humans. For example, the *arete* of the eye is seeing. Seeing is what makes an eye a good eye. If an eye cannot see, then we can say that it does not fulfill its function in the body. Furthermore, if we did not know what the eye is capable of, when it functions *well*, then we could not explain what it *is*. Then we would not know how to investigate the processes that go on in the eye. What the eye *is*, is defined in terms of what it does when it functions well (Brinkmann, 2004b). With regard to such functional entities as an eye, the fact-value dichotomy quickly breaks down, because any account of what such entities *are*, presupposes knowledge about what they *ought* to do. Concerning functional entities, it becomes quite easy to come up with counter examples to the idea of a naturalistic fallacy: From "he is a sea captain," for example, it seems logically valid to conclude that "he ought to do whatever a sea captain ought to do" (A.N. Prior's example, discussed in MacIntyre, 1985a:57). The sort of teleological explanation relevant to explain functional entities was exorcised from physics by Galileo and Newton, but defenders of the function argument believe that it is still necessary in the life sciences, including psychology.

An example of teleological explanation in psychology could be problem solving. Problem solving is a success term, defined in terms of its goal. It is difficult to recognize something as "problem solving" unless we understand what it means to *succeed* in solving problems. For otherwise, we could say that the person sitting in a chair watching football is engaged in problem solving, which *may* be the case, of course, but that judgment requires additional information in order to be justified. In short, the function argument says that when we pick out and study human functions (thinking, feeling, acting), we presuppose the existence of moral normativities, for without these we could not understand or identify functions as such.

The question is, however, how to characterize and ground human functions. Recently, there have been interesting efforts in evolutionary theory to argue against the naturalistic fallacy with reference to human functions, for example by Casebeer (2003) in his *Natural Ethical Facts*. Casebeer argues that "Hume's law" (no ought from is) and Moore's open-question argument ("good" is indefinable) both rely on a strict separation of analytical judgments (true or false in virtue of word meanings) and synthetic judgments (true or false in virtue of empirical states of affair), which Quine (1951) effectively undermined as one of the unjustified dogmas of empiricism. This is not the place to go into Quine's difficult and controversial argument (see Brinkmann, 2005), for what is interesting about Casebeer's project is his attempt to naturalize values and yet preserve their essential normativity. Unlike Hume, who held a non-cognitivist naturalist position about moral claims, implying that moral judgments cannot be true or false (for the reason that there is nothing in the world that could make them true, a view that is also defended by Ruse, 1991) – Casebeer argues in favor of cognitivist naturalism (moral judgments concern natural facts and thus *can* be true or false).

What facts, then, make moral judgments true? Casebeer here follows Aristotle and argues that functional facts about human beings fully fix normative claims (Casebeer, 2003:4). That is, moral facts are functional facts, facts about the proper functions of human beings. Functions are defined as dispositions and powers that explain the maintenance of a trait in a selective context (p. 52). We thus explain that the function of X is Z by saying that X is there because it does Z. For example, there are scissors because humans needed to cut things, and a scissor that is unable to cut is dysfunctional (at once a descriptive and evaluative claim). Analogously, to simplify one of Casebeer's examples, we are justified in saying that "ignoring the plight of those less fortunate than us can be ruled out as dysfunctional (in the naturalized Aristotelian sense)" (p. 63). For those who often ignore others will tend not to enter into productive social relationships, which is an important basic human function. Thus in Casebeer's view, it is a natural ethical fact that one ought not to ignore the plight of those less fortunate than us.

In contrast to Casebeer's biological account of functions as the objective foundation of ethics, there is another branch of Aristotelian philosophy that seeks to ground the objectivity of ethics in the ontology of practices (Davydova & Sharrock, 2003). The latter view is represented most forcefully in MacIntyre's (1985a) *After Virtue*, which we return to in detail in Chap. 7. MacIntyre appeals to practices and their internal goods as a way of overcoming the naturalistic fallacy. One is thus a good farmer if one performs farming well, which, in a way, is a different way of stating the conclusion of the constitutive rule argument, but, in MacIntyre's version, it is not "rules" that do the job of providing norms, but rather substantive values, that is, the kinds of activity that the proficient practitioner of some practice will display. In the conclusion, I shall return to the difference between the biological and the culturalist versions of the function argument.

The Thick Description Argument

I now come to the final line of argument, which I shall refer to as the thick description argument. This argument is connected to the approach in moral philosophy in the twentieth century that has been concerned with "thick ethical concepts." The notion of thick description goes back to two papers by Ryle (1971a, b), and was taken up a few years later by the anthropologist Geertz (1973). In moral philosophy, the idea of thick concepts has been developed in contrast to Kantian and utilitarian theories, concerned with "thin" notions of "right," "good," and "just." Some philosophers have called for "thicker" approaches to moral life (McDowell, 1981; Murdoch, 1997c; Taylor, 1989; Williams, 1985). It could perhaps reasonably be argued that "right," "good" and "just" are concepts without factual content; nothing more than a positive expression on behalf of the speaker, as the emotivists would have said. The emotivist argument runs into problems, however, when we consider such concepts as "cruel," "courageous," "brutal," "gentle," "loving," or "friendly." These concepts are thick ethical concepts, because they presuppose an understanding of contextual meaning, and, unlike thin concepts, they express a union of fact and value, as I shall now argue.

Almost all our words used to describe human action are thick ethical concepts (the following sections are based on Brinkmann, 2005). We have a few words to describe pure physical movement, of course, but at the level of human action and intention, facts and values are inseparable. We can exemplify this with the concept "cruel." This is clearly a value-laden word, for it would be contradictory to say "He is a very cruel person and a good man" (Putnam, 2002:34). But "cruel" is also a descriptive word. There are certain actions that are objectively and truly cruel, which is seen from the fact that the concept can be *correctly* (and incorrectly) applied. In normal circumstances, it would be incorrect to claim that brushing my teeth was a cruel act (of course, we can always a imagine contextual conditions that would make any given action cruel, e.g., if I brush my teeth as a way to ignore someone's need for help, but then it is actually my act of ignoring the other that is cruel).

Further, as argued by Williams (1985) and Taylor (1989), any attempt to find purely descriptive words, with which to describe a cruel action, fails. For example, if one tries to describe the pure physical movements of a torturer without including the moral qualities of the event, then the subject matter of the description changes, and it is no longer the same action that is described. In order to understand the qualities of the *action*, rather than the physiology of the movements, value-laden concepts are inevitable. As Louch described the difference between actions and movements: "Actions are movements seen and identified as warranted or not by circumstances. Movements, in a sense opposed to actions, are events seen as instances of interaction-push-pull, contact, collision" (1966:142–143). We do not first identify something as an action, and then look for situational reasons that would justify it. Rather, we initially "see" (as also proponents of the phenomenological argument would say) what warrants some behavior, thereby classifying it as an action, which means that identification and explanation of an action is often simultaneous. And all of this is possible because of thick ethical concepts: "description and explanation of human action is only possible by means of moral categories" (p. 21). Such moral categories, argued Louch, are indispensable if we – as psychologists and laypersons – want to understand central features of human action. Again, the thesis of the naturalistic fallacy seems unduly sterile and unhelpful if we want to understand the particulars of human action.

The main objection to the idea that thick ethical concepts can refer to objective features of the world is that it is possible to separate these concepts into a descriptive component and an "attitudinal" component, where only the former is said to refer to matters of fact, while the latter merely expresses a subjective attitude (Putnam, 2002:36). This counter argument fails, I believe, for it is impossible to say what the "purely descriptive" meaning of a word like "cruel" is, without using the word itself, or a similarly evaluative synonym (p. 38). The reader may try for himself or herself and find that "cruel" does not, for example, mean "causes deep pain" (which is closer to being purely descriptive, since doctors sometimes inflict deep pain on someone for the sake of that person's larger good). Even behaviors that do not cause obvious pain may nonetheless be cruel (e.g., if someone is prevented, without knowing it, from fulfilling his or her talent). When we rightly say that some action was cruel, then there is no way of picking out the same features of that action without using a synonymous thick ethical concept. As Taylor says, a thick ethical concept consequently picks out *real* features of the world, and the kind of value that is involved cannot be accounted for as subjective (Taylor, 1989:59). Knowing why actions x and y can both be described as "cruel" amounts to knowing *something about these actions* (rather than something about one's own attitudes). Thick ethical concepts pick out what Taylor calls "strongly valued goods," which are values that are not simply good because they are desired, but which are normative for desire; that is, they provide us with good (moral) reasons to be desired, whether or not one actually desires them. I return to this in the Chap. 6.

To sum up: My point has been that value-terms pick out features of the world (moral features) and that we could not understand ourselves, other people, or human action in general without them. They are indispensable to us, and we should therefore not try to remove them or purify them in psychological studies of human life, for they are an essential part of the language we need to act and live, and, as Taylor asks "what ought to trump the language in which I actually live my life?" (Taylor, 1989:58).

Conclusion

With the four arguments above, I have tried to show that in understanding a range of psychological phenomena, that is in understanding what human beings are capable of (perceiving oughtness, living together in institutional contexts, performing functions, and applying concepts to identify what people are doing), we constantly commit the naturalistic fallacy in the sense that we necessarily run together descriptive and evaluative characterizations. This is not because we are insufficiently enlightened, but is – if the arguments are valid – due to the nature of psychological phenomena. These differ from other sorts of phenomena (e.g., physiological ones) in being intrinsically and potentially subject to moral evaluation. For any psychological process we can ask not just "did it happen?" but also "was it done *well*?" There are normatively correct (and incorrect) patterns of thinking, feeling, and acting. Before moving on to a deeper discussion of these patterns – which, in my view, represent the basic fabric of psychological life – in the following chapters, I shall briefly compare, integrate, and discuss the different argumentative lines that have been addressed.

An appeal to the naturalistic fallacy is an appeal to the view that moral terms cannot be defined in non-moral terms – either because one believes that the referents of moral terms are irreducibly *sui generis* (cf. Moore) or because one finds that there really are no such referents (cf. Hume). Some of the critics of the naturalistic fallacy referred to above can be characterized as naturalists proper, in the sense that they believe that it is indeed possible to define normative moral terms in a non-moral language (e.g., Casebeer, 2003, in referring to evolved functions). Others, like MacIntyre (1985a) or Harré (2004), are not so much concerned with the issue of defining moral terms in non-moral terms, as they are with pointing out that, in psychology, normativity seems to go all the way down, so to speak. Psychological phenomena begin when normativity begins, that is, when there is a difference between correct and incorrect (perception, remembering, problem solving, etc.). The difference between these accounts can be understood in terms of different emphases on our first and second natures, respectively. Is the objective foundation of ethics to be understood in terms of "first nature", that is, our biological functions and needs, or is it to be understood in terms of "second nature", that is, our capacity to acquire habits through socialization and participate in social practices (Reader, 2000)? From where does the normativity in human lives arise?

This discussion needs to be brought to the attention of psychologists, for it is perhaps the most important discussion for those who are skeptical of the naturalistic fallacy and a strict fact-value dichotomy. As we shall see in the Chap. 6 and most directly in Chap. 7, I agree with the philosopher of science, Rouse (2007), that normativity, that is, those patterns that give us standards of correctness, are most fruitfully thought of as embedded in social practices. Normativity, by my lights, demands the existence of practices, that is, temporally extended ways of doing things, achieving goals, and cooperating, and there is no easy way from first nature biological accounts to a theory of practices. Furthermore, if first nature accounts, such as Casebeer's (2003), implies the view that evolved functions are what give us moral reasons for action, they then face the problem that we normally define moral reasons with reference to *objects* and *events* rather than functions. For example, if I say that I am acting morally for the reason that this action will express a basic human function, then a justified objection could be that this seems to disqualify the action as a moral one (since a moral action is hardly done for the sake of some function, but for the sake of alleviating someone's pain, for example). It would be more natural to say that a function (or a virtue) can enable us "to discern and act on good reasons, than to say that it can constitute those reasons" (Reader, 2000:361). And reasons are best thought of as located in the moral order of social practices (Brinkmann, 2006b).

The four arguments above can be seen as converging on the fundamental issue of practices. The phenomenological argument stressed the point that we can perceive meanings, values, and reasons for action, but these can be picked out only on the background of practices. Following Wittgenstein (1953), we should think of meaning as answerable to public criteria in social practices. The meaning of a word, a gesture, or an action, is not anything private in a subjective, inner realm, but is a use in shared social practice: "The meaning of anything in the domain of human consciousness is revealed by asking what role it plays in some human practice" (Harré, 2004:6). The constitutive rule argument showed that human institutions or practices are constituted by rules, which ground a fundamental normativity in social life, and a similar line of thinking lay behind the function and thick description arguments.

The question of the implications of this conclusion for psychology is difficult to answer. I believe that two things can be said, however. First, that psychologists ought to focus more on what their more or less implicit evaluative standpoints mean for how and what they study, and how they may come to affect what they study.

Those who study the intentional and normative world of psychological phenomena are themselves part of that world and may come to change it, as we saw in Part I. As John Shotter has put it, we should recognize that

> all our talk about human conduct is *normative*, that is, that in anything intelligible we say about it, we presuppose *judgments* as to whether it is right or wrong, fitting or unfitting, appropriate or inappropriate, successful or unsuccessful, etc., judgments which themselves may be grounded in reasons that in turn may be evaluated as good or bad reasons (Shotter, 1999:31).

Second, if psychological phenomena are at bottom normative and laden with moral value, and if we cannot but commit the naturalistic fallacy in describing them, then successful understanding of such phenomena depends on the character of those who understand. If so, it means that practitioners of moral science, among them psychologists, should themselves be moral practitioners – not just in the sense of investigating phenomena with irreducible moral properties, but also in the sense of needing a moral outlook in order to fully understand their phenomena.

Chapter 6
Moral Realism

Introduction: Moral Reality

Moral realism is usually presented as the view that there is such a thing as moral knowledge or moral cognition (for that reason, moral realism is conventionally referred to as moral "cognitivism" in the philosophical literature, but since this term has quite different meanings in psychology, I shall stick to "realism"). Nagel (1986:139) defines moral realism as "the view that propositions about what gives us reasons for action can be true or false independently of how things appear to us." It may appear to me that I ought to leave my family and travel to India in order to find and realize my true self. If one is a moral realist, then, according to Nagel's definition, it is either true or false *independently* of how the situation subjectively appears to me that I have a good reason to go to India. There is a fact of the matter, which transcends my subjective perspective. If my reason is a good reason, then it is good, not because it *appears* good or *feels* right to me, but because it hooks up with certain moral values that are there independently of my personal feelings, preferences, and attitudes. If, on the contrary, the reason is a bad reason, then it is likewise bad regardless of how I may subjectively feel about it. What is a good reason is not up to me, but is a matter of the structure of the space of reasons. Often, our reasons and evaluations are implicit in our reactions, feelings, and intuitions – in our "moral know-how" (Laitinen, 2002b) – rather than in formulated moral views and theories. But like explicit propositions, these reactions, feelings, and intuitions can still be more or less correct and morally legitimate, according to the moral realist. The realist claim is that what is morally good and right is not good and right because someone happens to like it. Rather, we *ought* to like what is morally good, and likewise detest what is morally bad. Indeed, we have *moral reasons* to like what is morally good, and detest what is morally bad. This is the view that I shall try to defend.

A defining aspect of what I called the psychological social imaginary in Part I is the view that something is morally good or bad because of psychological factors (psychologism). Such psychological factors can be feelings, preferences, or de facto psychological modes of reasoning. It is quite common for moral *anti-realists*, that is, for those who disagree with moral realists that there are moral reasons that

S. Brinkmann, *Psychology as a Moral Science: Perspectives on Normativity*,
DOI 10.1007/978-1-4419-7067-1_6, © Springer Science+Business Media, LLC 2011

stand or fall independently of the agent's subjective perspective, to emphasize such psychological factors as the crux of morality, and thereby *psychologize* morality. In logic, as we have seen, psychologism is illegitimate. It is not our mental operations that are the source of the validity of logical inference, for logic is the normative standard in light of which we can evaluate the validity of psychological thought processes. Psychologists have shown, for example, that when people are given the following pieces of information: "A bat and a ball cost $1.10 in total; the bat costs $1 more than the ball" and asked "how much does the ball cost?," most people give a wrong answer: 10 cents. This is wrong, because the bat would then cost $1.10, and both would cost $1.20 in total (the correct answer is thus 5 cents).

The point is that even if the thought processes of the large majority of the population prompt them to give a certain answer, in itself, this gives us insufficient reason to take for granted that the answer is correct. There are good and bad ways of inference in logic that stand or fall independently of how individuals happen to reason logically, and therefore the normativity of logic cannot be reduced to psychological functioning. A view like this is a realist view concerning logical reasoning. The moral realist's claim is that something similar is the case with regard to moral normativity: this too cannot be reduced to psychological functioning, for morality is the normative standard in light of which we evaluate our psychological functioning, for example perceptions, actions, and emotions. As MacIntyre (1988) has argued (cf. the introduction to this book), a normative evaluative background is needed in order for us to recognize something as a distinct psychological phenomenon in the first place, so it cannot be the case that such a background can be reduced to psychological functioning.

Anti-Anti-Realism

Moral realism is in some ways an odd term. Although Aristotle's "moral realism" in many ways inspires the kind of moral realism developed in this study, it seems forced to talk about Aristotle as a moral realist. *We* can indeed understand Aristotle as a moral realist, as many philosophers do (e.g., Lovibond, 1995), but we should bear in mind that in his own time, his form of realism was unproblematic in a sense that is unattainable to us moderners. As far as I know, Aristotle nowhere sets out to *prove* that there is a real and non-reducible moral realm, as some of today's moral theorists do. Indeed, he did not even have the modern idea of ethics or moral philosophy as a distinct field of inquiry. Of course, his major work, in what we refer to as ethics or moral philosophy today, comes down to us as the *Nichomachean Ethics*, but "ethics" here means something different from how we use the term today. It is better translated as "pertaining to character." The *Ethics* is about the proper development of character, the perfection of the human being, and could be classified as normative personality psychology. All ethics were "virtue ethics" in Greek philosophy (MacIntyre, 2001), because all ethics concerned the proper development of human excellences or virtues (*arete*).

Sabina Lovibond has aptly characterized the anti-realist position in moral philosophy:

> Moral judgments, it is claimed, lack truth-status – they are not the sort of utterance which can be true or false – because there is nothing in the world which *makes* them true, in the way that the physical condition of the world makes remarks about material objects true (Lovibond, 1983:1).

Modern moral realism has been an attempt to show that this is wrong; that there is in fact something in the world that can make moral utterances true or false. But in this light, moral realism is a derived position. It has largely been formulated, not positively on its own unproblematic terms in the way Aristotle had developed his empirical ethics (Aristotle's ethics could be "empirical" because he was never in doubt that the universe in which he lived had observable properties of value), but as a negative project in *opposition* to the kinds of subjectivism, emotivism, and error-theory that reigned in twentieth-century philosophy and social science.[1]

My point is that although I aim to develop and defend a version of moral realism that is relevant to psychology, my position is perhaps better characterized as *anti-anti-realism*. This is how a leading Wittgensteinian "moral realist," John McDowell, characterizes his own work. He says that anti-realist positions (e.g., subjectivism, emotivism, the error-theory) "are responses to a misconception of the significance of the obvious fact that ethical, and more generally evaluative, thinking is not science" (McDowell, 1998:viii). This misconception is crystallized in the view that if it is not science, then it cannot be objective. But this is simply a prejudice. In contrast to this form of scientistic dogmatism, I shall argue, not that we need to introduce morals and values into our world de novo, but that such values are *already* there as an indispensable and inescapable part of human mental life. We simply stand in need of being reminded of this common-sense point, and this is a task that is undertaken in the present chapter.

A Transcendental Argument

My argument rests on two central pillars: it appeals to *human experience*, and it is, in the last resort, an argument ad hominem (meaning "against the man"). These two features can be said to make it a transcendental argument (Holiday, 1988). Transcendental arguments are traditionally associated with the name of Immanuel Kant. Kant devised his transcendental argument in the first *Critique* to establish the necessary conditions for conceptualized experience. The starting point was that we do in fact have experience, and the transcendental argument was intended to give an account of its possibility. However, the appeal to experience in what follows is

---

[1] *Subjectivism* is the view that ideas about the good are subjective; *emotivism* is the idea that moral utterances are pure emotional exclamations; and the *error-theory* advocated by Mackie (1977) is the view that we may *think* we can refer to objective moral values, but that we are simply in error in thinking so.

not primarily in debt to Kant's philosophy (except in the obvious sense that all transcendental arguments are), but rather to philosophers and psychologists of pragmatist, phenomenological, and ecological bends.

Besides being grounded in, or appealing to, human experience, the argument below is also ad hominem. Sometimes this is employed as a term of abuse, implying that an argument is not strictly logically valid, but *merely* ad hominem. However, if we agree with Aristotle that ethics, along with politics, is among the *practical sciences*, then "we are studying not to know what goodness is, but how to become good men" (Aristotle, 1976:93). It can be argued that it makes little sense to be interested in morality, if one is not at the same time interested in acting better, or in improving the human condition. Similarly, it makes little or no sense to want to know logic if one does not want to learn how to reason logically, for what it means to "know" logic, is to be able to perform logical operations. If Aristotle was right that moral knowledge is practical knowledge, then the same applies here: What it means to "know," morally speaking, is to be able to perform moral actions. In this light, arguments ad hominem are very useful in answering the moral subjectivist or the moral skeptic. For, if successful, such arguments

> Seek to convict the skeptic of inconsistency on the grounds that his skeptical utterances do not square with the fact that he, being what he is, is uttering them. They try to show that, because of what he is, he has no *title* to doubt what he claims to doubt, no right to call what he does "doubting" (Holiday, 1988:162).

Successful arguments ad hominem thus show that what someone says contradicts what he already is or does. In this respect, they point out subtle performative self-contradictions (a simple example of a performative self-contradiction is if someone says "language does not exist!"). Holiday's (1988) own interesting contribution, to be addressed more fully below, aims to show that the very fact that we are language-using creatures is possible only because of a range of objective moral values presupposed in the practice of speaking (e.g., truthfulness and trust).

In the Thick of Moral Things

Can we ever, for example, as psychological theorists, be "outside" the sphere of morality? Not according to the view that I wish to defend. Any claim *about* morality (including those put forward by subjectivists and skeptics) must ultimately, in my view, be seen as *moral* claims. As Dewey once said: "There is and can be, then, no rigid line between 'ideas about morals' [...] and 'moral ideas.' The former are the latter in the making" (Dewey, 1891:196). Nagel also puts this well, when he refers to Ronald Dworkin's assertion: "the only way to answer skepticism, relativism, and subjectivism about morality is to meet it with first-order moral arguments. He [Dworkin] holds that the skeptical positions must themselves be understood as moral claims – that they are unintelligible as anything else" (Nagel, 1997:vii). Lovibond has advocated a similar view – she calls it the "practical reason" approach to ethics – that pictures the ethical theorist (and, I would add, the psychologist)

> Not as a mere onlooker, but also, and primarily, as an ethical subject, actively engaged in the mode of thought on which (as a theorist) she reflects, and finding her incentive to theory precisely in the desire to think more competently or intelligently about ethical matters (Lovibond, 2002:8).

This means that ethical theory becomes instrumental for ethical practice, as also Dewey repeatedly argued. Ethical theory has no point, purpose, or authority in itself, but gets its point, purpose, and authority from being able to assist us in living our everyday lives in responsible, just, caring, and satisfying ways. We (here included the moral subjectivist) are in the thick of moral things, the space of reasons, and there is no way to escape moral argument. Even the hard-nosed skeptic must provide us with *reasons* to believe him, and, if he wants to be taken seriously, he must also be willing to discuss with us whether his reasons for skepticism are *good* reasons, whether they are only reasons *for him* (i.e., purely subjective), whether they are reasonably well articulated and thought through, and he must also be willing to discuss whether the real-world consequences of his skeptical beliefs are desirable. All these points involve normative – indeed moral – values, and presuppose the entire range of human moral practices of giving (good) reasons for one's beliefs and actions, which is exactly what is at stake in the skeptic's project (namely to show that all such reasons are ultimately subjective).

To sum up: My point is that no claim *about* morality, neither those put forth by realists, nor those argued by skeptics, subjectivists or relativists, are immune to *moral* criticism. This view is close to what is sometimes called "quietism" in modern analytical philosophy. Quietism claims that any investigation of our linguistic and moral practices "must be carried out from a position of immanence within them; [it is] a policy of giving up that fantasized external standpoint from which we could supposedly pass judgment on whether this or that entire region of discourse succeeds in making contact with the 'real world'" (Lovibond, 2002:22). Terry Pinkard has argued that this approach to normativity is found in (Dewey's) pragmatism: From the normative point of view, he says, "there is no 'outside'; wherever we stand, we are always, to use a Sellarsian turn of phrase, inside the practice of giving and asking for reasons" (Pinkard, 2007:144). This is our fundamental moral *and* psychological reality.

There is no external standpoint outside morality from which to judge whether it (as an entire field of discourse and action) is real or not. The fact that we cannot do without it is pragmatic proof of its reality. As Charles Taylor says in his own version of the ad hominem argument against the moral skeptic: "What is real is what you have to deal with, what won't go away just because it doesn't fit with your prejudices" (Taylor, 1989:59). Morality is real, because it will not go away, and even the moral skeptic must, in his practical attempts at persuading us that he is right, take it for granted if he is interested in giving us reasons, and receiving them in return. Morality is not just "real" but also "objective" in this sense,[2] at least if Dewey was right in defining "objective" as "that which objects, that to which

[2] This addition (that morality is real *and* objective) is important, for illusions, for example, are also real (but subjective), and I want to avoid the conclusion of the error-theory that morality is a "real illusion."

frustration is due" (Dewey, 1925:239). A moral dilemma can be frustrating, because we cannot subjectively annul the objective values that make up the dilemma. The objective, Dewey argued, is that which *objects* to our dealings with it, and *resists* us. I shall argue that morality is objective in this sense.

In the remainder of this chapter, I will try to describe some of the ways in which we are inescapably in "the thick of moral things." I point to three psychological dimensions that are saturated with moral issues: (1) *Moral perception*: I first follow up on the preceding chapter by arguing that moral phenomena already appear on the level of our perception of the world, and that clear perception is sometimes a sufficient guide to moral action. (2) *Moral action and re-action*: I argue secondly that our ways of acting in the world, and *re*-acting to events, presuppose the reality of moral values. There is such a thing as *good reasons* for doing something, and I argue that our capacity for evaluating different courses of action ultimately makes sense only given certain irreducible moral values. In this context, I also analyze emotions as moral phenomena. (3) *Moral identity*: Third, and closely related to the second point, I argue that we cannot make sense of human identity without presupposing the reality of moral values. As Taylor has argued, humans cannot do without some orientation to the good (for Taylor, this is what it means to have an identity), and this good cannot be construed as something of our own making, for I can be a "someone" only because certain things matter to me morally, *prior* to reflection, choice, and personal desires. I end by addressing a form of reductionism that is complementary to the kind found in evolutionary psychology, namely social constructionist reductionism.

Moral Perception

As Blum has pointed out in his important contribution to the study of moral perception, moral inquiry has traditionally masked the importance of moral perception by focusing instead on action-guiding rules and principles, choice and decision, universality, impartiality, obligation, and right action (Blum, 1991:701). The focus has often exclusively been on the procedures to be followed in reaching the proper conclusion as to what constitutes moral action. But, as Blum makes clear:

> An agent may reason well in moral situations, uphold the strictest standards of impartiality for testing her maxims and moral principles, and be adept at deliberation. Yet unless she perceives moral situations as moral situations, and unless she perceives their moral character accurately, her moral principles and skill at deliberation will be for naught and may even lead her astray. In fact one of the most important moral differences between people is between those who miss and those who see various moral features of situations confronting them (Blum, 1991:701).

Moral perception, as Blum presents it, necessarily precedes reflective and deliberative modes of moral reasoning. However, it makes sense to perceive moral features accurately only if there are moral features to perceive. So Blum's account of moral perception presupposes the reality of moral phenomena; it presupposes a version of moral realism. As we saw in previous chapters, we are continually confronted with "phenomenological oughts" in our daily lives. We are able to experience "oughtness" directly. Here, I wish to expand on this by introducing the more radical claim that perception is often not just a *necessary* component in moral action, but also sometimes a *sufficient* component.

The Phenomenological "Ought"

Little evidence is needed to argue that in our everyday lives we do not operate in a value-neutral environment, but rather in a value-rich world that seems to call for appropriate patterns of moral response. We do not experience properties of value as something we subjectively "add" unto a naked physical world, but we are directly confronted with issues of value. We experience ourselves being *moved* by values that are there independently of our subjective perspective. To take a standard example: If I walk alongside a river and I suddenly notice someone in the water struggling for her or his life, I am directly confronted, not with a thing or a neutral event, but with a *demand*, namely. to try and help the person. This moral demand is phenomenologically prior to any deliberation I may engage in about the situation (indeed it can even be argued that it is *immoral* to suspend one's immediate inclination to help in this situation in order first to engage in moral reasoning). As Honneth says: "in normative contexts we act in an 'always already' perceptually disclosed world of moral facts" (Honneth, 2002:247), and we must therefore "really conceive of our moral knowledge as a perception of ethical states of affairs" (p. 254). Marginal traditions such as Gestalt psychologists, pragmatists, and ecological psychologists have all taken an interest in such perception.

Pragmatism and Ecological Psychology

The phenomenological descriptions of direct perception of "oughtness" found in Gestalt psychology, which were discussed in Chap. 5, have an important precursor in the writings of William James.[3] James's "radical empiricism" invites us to respect the qualities of everyday experience. "To be radical," James said, "an empiricism must neither admit into its constructions any element that is not directly experienced, nor exclude from them any element that is directly experienced" (James, 1912:42). James found that we not only experience objects, but also

[3] Herzog (1995) has argued that James in fact deeply influenced the phenomenology of Edmund Husserl from the outset. Husserl's copies of James's books are full of comments, notes, and translations. For me, the issue is not about which label is more adequate – be it phenomenology, pragmatism, Gestalt theory, radical empiricism, or ecological psychology – but their shared insistence on the richness of concrete human experience.

relations between objects. Therefore, he argued, such relations and "oughtnesses" should be respected by any inquiry into human experience. In fact, we do normally *not* experience one discrete object after another (unless we participate in some psychological experiment), but rather events, situations, connections, relations, meanings, and purposes. We experience situations calling for appropriate patterns of response. The world is not a huge collection of things in motion – what James pejoratively referred to as a "block universe" – but a world of events with many kinds of qualities that become and change, and are constantly in the making. The only reality that is available to us is "practical reality" in James's (1950:295) words, and, in our practical reality, we encounter things and situations in terms of what they are good for, in terms of their meanings, and "esthetic and moral experience reveal traits of real things as truly as does intellectual experience," as Dewey said (1925:19). In our practical reality, "things are objects to be treated, used, acted upon and with, enjoyed and endured, even more than things to be known. They are things *had* before they are things cognized" (p. 21). Pragmatists like James and Dewey presented moral values as essentially "adverbial," that is, qualities of *how* things are done, and argued that in evaluating how things are done, we can make objective moral judgments.

In his book on the development of ecological psychology, Harry Heft argues that "the conceptual underpinnings of James Gibson's ecological approach are traceable to James's later philosophical position" (Heft, 2001:16). Heft believes that James's "radical empiricism can be employed as a philosophical foundation for an ecological psychology that extends beyond the purview of Gibson's work" (p. 16). Along with Gestalt psychology and pragmatism, Gibson's ecological approach represents one of the few traditions in psychology that aim to respect concrete human experience by rightly placing values in the world. Gibson argued that the meanings or values of things – affordances – in the environment can be "directly perceived" (Gibson, 1986:127). He defined affordances of the environment as "what it *offers* the animal, what it *provides* or *furnishes*, either for good or ill" (p. 127). It is affordances that we primarily perceive when we encounter objects. They do not *cause* behavior, but *afford* kinds of behavior (and are thus *normative*). We encounter objects as graspable, movable, sit-on-able, and so on, before we encounter them as discrete physical entities or as elemental sense data.

Still and Good (1998) have extended the Gibsonian framework to encompass more wide-ranging psychological issues, such as moral responsibility. They argue that Gibson's view of direct perception can favorably be coupled with the moral phenomenology of Emmanuel Lévinas. As they say, ecological information specifies ethical demands, just as it specifies morally neutral affordances like a chair's sit-on-ableness. In human encounters we can, if we have been provided with adequate moral habits in our upbringing, directly perceive the ethical responsibilities we have toward others, and such perception is unmediated by knowledge, rules, and theories. Also Hodges and Baron (1992) have developed Gibson's ideas further by focusing on values as embedded in the "environmental arrays" and activities of perceivers. They understand perception as an "achievement term" in that it is value-realizing.

The Value-Realizing Nature of Perceptual Activity

What they mean by this – that perception is value-realizing – can be understood by drawing in Gibson's and Dewey's analyses of perception as a function of action. Costall (2004b) has argued that Gibson's final work should rightly be understood as a theory of *agency* rather than perception. Perception is not a passive mirroring of a static external reality, but is a function of our moving around in a changing world where we examine objects, do things, and have intentions that we try to realize.

Dewey understood perception as a function of underlying habits. In Dewey's early and seminal critique of stimulus–response psychologies, he demonstrated that there is no way to distinguish between stimulus and response in absolute terms, but that both should be understood as aspects of the organism's primary ongoing activity in a meaningful world (Dewey, 1896). Nothing simply *is* a stimulus, but something can *become* a stimulus if the organism is able to pick it out as such on the background of appropriate habits. The green light is only a stimulus to get one's car going if one has acquired the habit of driving in the traffic of modern society. Noticing the stimulus of the green light is an end-point (that is likely to prompt further action, of course), an achievement of perceptual activity inherent in the social practice of driving cars, rather than a pure beginning in a stimulus–response schema. The search for a stimulus is thus a search for the *value* of the object (Backe, 2001). Engaged in the activity of driving my car from work, perceiving the affordance of the green light is a value-realizing activity, since it does some good in relation to my larger activity of getting home. This is the meaning of perception as a value-realizing activity based on habits or practices.

Dewey's analysis of habitual activity is very rich, and can even be said to form the basis for his entire psychology. In his framework, we "know" with our habits, not with our "consciousness" (Dewey, 1922:182). Thus, "The reason a baby can know little and an experienced adult know much when confronting the same things [the same 'stimuli,' SB] is not because the latter has a 'mind' which the former has not, but because one has already formed habits which the other has still to acquire" (p. 182). The adult is able to perceive affordances and values of the environment, because he or she has acquired the relevant habits, different and fine-grained ways of dealing with the world and its objects. Dewey's pragmatic psychology and Gibson's ecological psychology here share the anti-cognitivist view that perception is not mediated by mental representations, but rather by attunement to specific relations among ways of acting in the world, that is, by habits (see Manicas, 2002:287). Habits are what enable us to navigate the space of reasons. Habits are "second-nature" (and enable us to understand the green light as a reason to get the car going).

Summing up, we have seen that phenomenological (i.e., Gestalt), pragmatist, and ecological perspectives in psychology all have tried to place values, including moral values, in the objective world, and have provided arguments to the effect that we are able to perceive value – "oughtness" – directly. As Blum has argued: "In a given situation moral perception comes on the scene prior to moral judgment;

moral perception can lead to moral action outside the operation of judgment entirely" (Blum, 1991:702). If I correctly and accurately perceive the situation – that someone is drowning unless I react – then I am provided with a reason to do something without having to engage in procedural moral thinking.

The Sufficiency of Perception

Does this not mean that accurate perception will remove entirely the need for moral reasoning, deliberation, and choice? Philosopher and novelist Iris Murdoch defended this view: "If I attend properly I will have no choices and this is the ultimate condition to be aimed at" (Murdoch, 1997b:331). The person with a well-formed moral character has no need to engage in moral reflection according to Murdoch, for she clearly *perceives* what to do. Levine (1998) has also argued that we can settle moral dilemmas by "describing particulars in a judgmental way" (p. 4). This means *seeing* the moral demands that inhere in situations and events. According to Aristotle, what is needed here is not scientific knowledge or abstractions, but practical wisdom or *phronesis*, which, as he said, "apprehends the ultimate particular, which cannot be apprehended by scientific knowledge, but only by perception" (Aristotle, 1976:215). What should be perceived is, in ecological terms, what situations afford morally. What situations afford morally is sometimes easy to see, as is the case if someone is drowning in a river, but quite often it is very difficult, and people then tend to disagree about it. As McDowell (1998:72) has stated: "ethical reality is immensely difficult to see clearly." Moral situations are indeed often vague and complex, and there is no reason why moral dilemmas should not be understood as genuinely real.[4] However, my claim is that we learn what to do in moral situations, including vague ones, by perceiving them accurately, including here perceiving their very vagueness, rather than detaching ourselves from them and employ prefigured procedures of moral decision-making.

One of Levine's instructive examples is sexual harassment. It is clear to anyone who understands this word that it denotes something that is morally wrong. If we identify some action as an instance of sexual harassment, then we are provided with a reason to intervene, to testify in court, to call for help, or something like that. The crucial question, then, is not whether sexual harassment is wrong or not, but whether some particular event can correctly be identified as an instance of it. And this is a matter of perceptual judgment. As Levine says, "descriptive words can carry strong moral connotations – indeed, they can make moral judgments all by themselves. 'Sexual harassment' is certainly one such term" (Levine, 1998:21). We may deny that the term applies in a given

[4] According to several dominant modern procedural moral theories, moral dilemmas are illusory. For Kant, if two duties contradict each other, one is not a duty, and utilitarians would say that there is always one course of action that maximizes preference satisfaction more than another. These traditions do not admit that *vagueness* can be a real element of moral or psychological reality.

case, but once we apply it to describe some act, we implicitly condemn the act morally. It is thus a "thick ethical concept" (cf. Chap. 5). Although the judgment as to whether something counts as sexual harassment or not can be difficult, such a particularistic perceptual judgment "often does the whole work of ethics" (p. 28) without the need for universal principles. As Nussbaum puts it plain and simple: "Principles are authoritative only insofar as they are correct; but they are correct only insofar as they do not err with regard to the particulars" (Nussbaum, 1990:69).[5]

On many traditional psychological accounts of "perception," it would seem odd to claim that we can *perceive* that something is sexual harassment. That judgment would seem to involve some "higher-order" psychological function (perhaps "memory," because one has to remember the rules for what counts as sexual harassment). And there is no doubt that one can perceive situations as instances of sexual harassment only if one masters the concept "sexual harassment" (an instance of Hacking's "human kinds"). But if we agree with phenomenology, pragmatism, and ecological psychology that "meanings ain't in the head" (as Putnam, 1973, famously put it), but rather in the world as aspects of human interaction, then it is less mysterious to claim that we can "perceive" that something is an act of sexual harassment. For perception here is already "judgmental," and indeed "evaluative." From birth, we learn to *see* and *respond* to moral properties by participating in practices, before we learn to give reflective accounts and justify our actions. We are "introduced to values by way of training, habits, and institutional influences" (Hatab, 1995:406), which is to say that we are already shaped by ethics before we come to reflect on it (p. 405). We are brought up as moral realists.

The preliminary conclusions are then, first, that accurate perception is sometimes sufficient to guarantee moral action, and, second, that moral perception seems to be of real properties of situations and events. This has to be so if we grant two things about the example of sexual harassment: that it is real (not something we subjectively choose to call so) and that it is morally wrong. A third conclusion that was touched upon was that perception must be understood as a function of action, which points forward to the section below on action.

Impressions and Perceptions

Against the objection that I have granted too much to perception (I have granted that we can perceive meanings, values, and affordances directly), let me point to Merleau-Ponty's phenomenology of perception, which tried to demonstrate that

[5] I do not, however, agree completely with Levine that perceptual judgment can do the whole work of ethics, for it seems to be the case that more reflective processes of reasoning become important when normal moral habits break down or become disturbed. Furthermore, it is important to bear in mind that perception always occurs as part of some practice (and it would be more accurate to speak consistently about perceptual *activity* instead of simply "perception"), which means that perception is always cultivated, cultural, and historical. Perception is mediated by habits and social practices (Brinkmann, 2004a).

"To perceive is not to experience a host of impressions accompanied by memories capable of clinching them; it is to see, standing forth from a cluster of data, an immanent significance without which no appeal to memory is possible" (Merleau-Ponty, 1945:26). I have argued that such "immanent significance" is imbued with value and moral requiredness. One of the great fallacies of those psychologies that do not begin with everyday experience is to grant perception too little; or, rather, to argue (like Hume) that perception consists of "sensation" or "impression" (which is allegedly completely elemental and neutral) plus memory (which, then, is conceived as a faculty that somehow "adds" value and meaning to naked sensory input). What we have learned from the marginal phenomenological, Gestalt, pragmatist, and ecological schools of psychology is that perception from the outset is able to disclose a rich world of values and meanings for us, and that it does so because it is *not* passive and spectator-like, but a function of the activities of an organism that *does* things in a changing world. What empiricists since Hume have called pure impression (i.e., sensation) turns out to be a myth:

> The pure impression is, therefore, not only indiscoverable, but also imperceptible and so inconceivable as an instant of perception. [...] An isolated datum of perception is inconceivable, at least if we do the mental experiment of attempting to perceive such a thing (Merleau-Ponty, 1945:4).

Attempting to isolate pure sense impressions in one's experience is indeed an artificial exercise, and perhaps an impossible exercise. To think that this is what experience is really made up of is to commit what James (1950) called the psychologist's fallacy: It confuses a reflective onlooker's or theoretician's view of some part of the world with how that part actually *is*. Committing this fallacy in relation to sense impression can have serious consequences for how we are able to conceptualize values in psychology, as Merleau-Ponty was well aware: "Every evaluation had to be the outcome of a transfer whereby complex situations became capable of awakening elementary impressions of pleasure and pain, impressions bound up, in turn, with nervous processes" (Merleau-Ponty, 1945:64). If one believes that our only contact with the world is through naked and elementary sense impressions, then the whole complex moral life of humans is reduced to pleasure–pain psychology, for simple pleasure and pain is all that is "awakened" by sense impressions, as Merleau-Ponty observes.

Evolution and Morality

In Chap. 5, I argued against the evolutionary perspective on morality in the context of a discussion of the naturalistic fallacy. I shall now expand on the critique of evolutionary psychology from the interpretive–pragmatic viewpoint. The evolutionary perspective on morality is of course part of a much broader tradition of evolutionary theory in psychology (Buss, 1999; Caporael, 2001; Tooby & Cosmides, 1992). The fundamental assumption of evolutionary psychology is that human psychological mechanisms are what they are because they have proved to be functionally adaptive, and thereby have been selected for in the course of natural history. What we call morality is likewise what it is, because it has been selected for. A radical version of the evolutionary approach was found in sociobiology. According to sociobiology, it is the continuing existence of the gene rather than the organism that plays the leading role in evolutionary history. Wilson thus claimed that morality has no other ultimate function than keeping genetic material intact (quoted from Midgley, 1994:5). Sociobiologists predict that altruistic actions will be more likely to take place if there is much, rather than little, genetic resemblance between the helper and the person being helped (Dawkins, 1986; Wilson, 1975). It is also predicted that people will form moral judgments that help them pass on their genes to the next generation (Wright, 1994). In this way, it can be explained why, as Wright seems to think, sexually very active women are morally more blameworthy than sexually very active men, since men rather than women have been created, biologically speaking, to spread their genes (p. 146).

One does not have to be professor of moral philosophy to recognize the fallacy lurking in this line of argument. Men also seem to be biologically prepared to die at a younger age than women, and yet it seems unreasonable to conclude that dying young is morally praiseworthy for men. I would question, in the first place, that there is this biological difference between men and women as regards sexual activity, but *even if it were the case*, this would give us no reason to conclude anything about its moral qualities. This demonstrates an important limitation in the sociobiological, indeed evolutionary, perspective: It cannot on its own terms account for why and how it identifies specific actions or traits as specifically moral. In itself, the fact that something is biologically or genetically hardwired gives us no reason to call it moral, for countless things are so hardwired without this tempting us to call them moral (e.g., the human preference for sweet foods).

Furthermore, as Gantt and Reber have pointed out, it borders on absurdity to want to confer the capacity for complex moral evaluation onto genes: In sociobiology "it seems to be simply assumed that genes are somehow able to do whatever it is necessary for them to do, no matter how abstracted from the actual chemical properties and processes the sociobiological account must become" (Gantt & Reber, 1999:20). Sociobiological accounts can be said to commit a category mistake in presenting genes, which operate at molecular levels, as having psychological skills like the capacity for value judgments and for developing "sophisticated social and behavioral 'stratagems' based on those judgments" (p. 20).[6]

The Basic Building Blocks of Morality?

When evolutionary psychologists study moral behavior in primates and try to find the basic "building blocks" of morality there (Flack & de Waal, 2000), they are hard pressed to articulate the basis on which they identify different kinds of primate

[6] To give an example, evolutionary psychologist David Barash claims that "Parenting is just a special case of genes looking out for themselves in the bodies of a special type of individual called offspring" (Barash, 1982:70).

behavior as moral. That something has survival value is not enough, for many a-moral things have survival value (e.g., being born with legs). Furthermore, if moral reactions were like genetically inbuilt instincts, as the evolutionary account seems to argue, they would be on a par with our reactions to rancid foods and rotten smells. Moral reactions would thus be nothing more than, and hence reducible to, visceral reactions. Something would be morally good because I generally react favorably to it, instead of the other way around; that we *ought* to respond with respect, love, and care to morally valuable properties. It would be impossible to find an ontological difference between my aversion to the Holocaust and my nausea at the smell of rotten fish. But, as Taylor (1989) argues, there is an irreducible difference related to the fact that in the latter case, there is nothing to articulate about my nausea, whereas in the former case, I can articulate why my reaction is *justified.*[7] The visceral reaction to rotten fish is neither justified nor unjustified; it is just there as a brute causal result, most likely as an inbuilt part of our olfactory sensory system (and is thus placed in "the space of causation").

Moral reactions are not just brute reactions but "also implicit acknowledgements of claims concerning their objects" (Taylor, 1989:7). They relate to claims about dignity, justice, respect, love, etc. When we articulate our moral instincts, we give what Taylor calls ontological accounts (p. 8). We cannot even *understand* the difference between visceral reactions and moral reactions unless we have an idea of moral ontology, an inescapable moral framework, which forms the evaluative background that gives meaning to our moral reactions and articulations of these. The fact that we do, in our everyday lives, understand the difference between pure visceral reactions and moral reactions is a powerful argument to the effect that there is an (ontological) difference between moral and visceral reactions that the evolutionary perspective does not take into account. Moral reactions refer to moral properties of objects and situations whereas visceral, instinctual reactions merely involve some physical and causal effect on our sensory system. The former unlike the latter refers to moral *meaning* in "the space of reasons."

The very attempt to find the basic "building blocks" of morality (Flack & de Waal, 2000) seems misguided from the interpretive–pragmatic perspective advocated in this book, for it always makes sense to ask: Is this basic "building block" really good? What good does it *do*? Good is, as Iris Murdoch argued, a sovereign concept, which we cannot understand via other concepts:

> Asking what Good is not like asking what Truth is or what Courage is, since, in explaining the latter, the idea of Good must enter in, it is that in the light of which the explanation must proceed [...] if we try to define Good as X we have to add that we mean of course a good X (Murdoch, 1997c:380).

We can thus ask what it is, for example, that allows evolutionary ethicists to claim that *quid pro quo* (one must give in order to get) is a *moral* (rather than a strategic)

[7] This parallels the argument I gave in the introduction that psychological phenomena qua psychological are normative: In relation to psychological phenomena there is potentially something to articulate because such phenomena exist in the space of reasons.

principle (cf. Høgh-Olesen, 2004)? The moral realist would insist that the answer would have to include, if it is to be morally illuminating, reference to moral values that are there *independently* of this evolutionary-based behavioral pattern, for it is not true that "good" simply means "in accordance with the principle of *quid pro quo*" or "whatever has survival value."

A Practical Reason Rejoinder

Furthermore, from a practical view of things, we should be suspicious of a theory of morality that instructs us to live as if it is not true! The evolutionary psychologist's belief that morality is a functional fiction does not just go contrary to everyday moral experience, but would also stifle human action were we to *act* on this belief. Along with Aristotle, I believe that it must be the task of a "moral theory" to help us act morally, to improve the human condition, and this requirement is hardly met by declaring all human experience of moral issues a functional fiction.

Even the starkest evolutionary account of morality does not escape our moral space, our practical reality, or the space of moral reasons. As Taylor has said, the evolutionary perspective is "motivated itself by moral reasons, and these reasons form an essential part of the picture of the frameworks people live by in our day" (Taylor, 1989:23). The moral reasons motivating current evolutionary (and other scientistic) accounts of morality typically involve the values of objectivity, truth, seeing the world without personal bias, and even sometimes a certain existentialist respect for humans who are able to look the absurdity of a disenchanted universe in the eye. Sociobiological psychologist David Barash (2000:1012) thus says that "existential philosophy and evolutionary biology enjoy substantial and hitherto unappreciated similarities – in particular an understanding of life's fundamental absurdity, a relentless and productive focus on the individual, and an optimistic presumption of freedom"; sociobiology thus "leaves us free at last to pursue our own, chosen purposes" (p. 1013) (one is reminded here of the psychological social imaginary as articulated by Carl Rogers).

Summing up the evolutionary challenge to moral realism and the idea that we can perceive objective moral properties, we can say that although the evolutionary theory is a useful resource for understanding our place as a species in a changing world, it cannot tell us the whole story; at least not about human morality and practical reason. This does not mean that morality is not a natural phenomenon. On the contrary, I believe that it *is* natural, real, and objective, but that the way it naturally exists, so to speak, is as a normative phenomenon (or as a range of normative phenomena in the plural). By declaring it a functional fiction, or an adaptive mechanism designed to keep genetic material intact, one radically reduces its normativity away. In fact, many evolutionary accounts of morality *psychologize* morality when seeing it as rooted in pure psychological functioning,[8] for all that can ever drive moral behavior on its account are psychological factors like preferences and desires, grounded in our

[8] Although not all: Midgley (1994) is one important exception.

genetic material. It is no coincidence that David Hume is the favored father figure for many evolutionary ethicists (see, e.g., Ruse, 1991:508).

Moral Action and Re-Action

We now move from perception to action. I take it as a fact of the psychology of action that in deciding how to act, we are sometimes able to evaluate whether our desires and preferences are *worth* realizing. There is such a thing as *good reasons* for doing something, and I will argue that our capacity for evaluating different courses of action makes sense only given certain irreducible moral values. Of course, I do not want to deny that we often act strategically and in immoral ways, but I maintain that we also have the capacity for moral action.

In Chap. 5 I argued against the view that the "thick ethical concepts" we use to describe and evaluate actions can be bifurcated into a purely descriptive and a purely evaluative part, and here I shall briefly expand on that argument. Following Wittgenstein (1953), we can say that understanding a word means being able to use it correctly in public practices, contexts, or what Wittgenstein called language games.[9] There is no other way to account for linguistic meaning than as use. The alternative would imply recourse to some private ostensive definition or mental content, but Wittgenstein's argument against the possibility of a logically private language has effectively barred this road.[10] The possession of a concept is not the possession of a mental image or a psychic representation, but is being able to *do* certain things (Bem & Looren de Jong, 1997:143).

Understanding a concept thus means being able to "go on" with the concept in public usage, in shared social practices. It involves being able to apply the concept correctly in relevant cases, and this also goes for thick ethical concepts such as "brutal," "courageous," or "cruel." As we saw in Chap. 5 there is no other way of adequately understanding and describing the cruelty of a cruel action, than by including the evaluative word "cruel" or a similarly evaluative synonym. Thus, we have an argument to the effect that values exist as part of the world. A genuine value term is no subjective projection, and if we master thick ethical concepts, we will often know what to do in a given situation, that is, we will know what we *ought* to

[9] And the interpretive–pragmatic view on morality argues similarly that understanding morality means being able to practice it correctly.

[10] A private language is a language that refers to what the speaker alone can know, that is, the private contents of one's mind, for example in the form of Humean impressions and ideas or mental representations in current psychological terminology. Wittgenstein rejected the idea that language can get its meaning from referring to private representations because if that were so, only the speaker herself could know if she used a word correctly, and there would thus be no difference between correct and incorrect usage, which means that language as such would be impossible (because *normativity* would be impossible). The fact that language *is* possible shows, Wittgenstein argues, that it does not derive its meaning from private concepts, definitions, representations, or ideas, but from public use in socially shared language games.

do in this situation.[11] Thick ethical concepts thus counter the alleged "naturalistic fallacy," as we saw. For a range of concepts we cannot describe adequately how the world *is* without implying what we *ought* to do.

Strongly Valued Goods

Thick ethical description and concepts are related to what Charles Taylor has called strong evaluations, that is, qualitative distinctions concerning the worth of different options. Strong evaluations refer to strongly valued goods, or simply moral values, and I will argue that such moral values are real, and a transcendental condition for the existence of human action as we know it.

It has long been argued by a minority group in analytical philosophy of action that "observation, description and explanation of human action is only possible by means of moral categories" (Louch, 1966:21). The argument says that "when we offer explanations of human behavior, we are seeing that behavior as justified by the circumstances in which it occurs" (p. 26). According to Alfred Louch, this means that explanation of human action is moral explanation, for we can only identify actions by describing experience by means of moral concepts (pp. 26–27). This will also be my argument. But what does it mean? It means that we have an ability to distinguish between *behavior* that is *caused*, such as my trembling in cold weather, and *action*, which is identified in terms of *reasons* calling for justification. When we explain ourselves to others and answer the questions "What did you do?" and "Why did you do that?," we offer descriptions and justifications in the light of our motives in the situation (we place the episode in the space of reasons). Having a motive implies "that the situation in which the person finds himself would justify him acting in a certain way" (p. 79).

Sometimes we explain ourselves by saying "I did this, because I wanted to!"[12] To explain and deliberate about one's actions in light of one's desires and preferences ("what do I want the most?"), is what Taylor (1985d) calls weak evaluation. A paradigm example of weak evaluation is when we go to a restaurant and try to determine which meal to order. One scans the menu and finally decides to go for the fish. When asked the reason why, there is rarely anything more to say than "because

[11] One of the oldest philosophical problems, still unsolved, is the problem of weakness of the will – *akrasia* – also discussed in depth by Aristotle. It arises also in the context of my account, because I need to explain the fact that someone can correctly identify the moral properties of a situation (thus being guided by the world) and yet fail to act properly (fail to follow the action guiding element). If moral judgements are *practical*, that is, express themselves in action, and *objective*, that is, refer to moral facts, then we should not be able to perceive moral facts without acting on them. Internalism is the view that moral judgments are indeed intrinsically motivating, and I believe that Aristotle's (1976) and Lovibond's (2002) defense of a modified internalism is promising, although I will not go further into this age-old debate in this book.

[12] Although, as Louch (1966) has demonstrated, such explanation is vacuous (and a likely reaction is: "Of course you want to – otherwise you wouldn't be doing it!"). The only context in which that explanation is informative is when others may not have expected the action or suspect that it was done under pressure (p. 93).

I felt like it!" One weighs one's desires, and determines which is the quantitatively strongest one. In weak evaluation, one's motivations, desires, and preferences are put on a par. They are considered as homogenous and put on the same scale. But if the model of weak evaluation were the only model a person had recourse to in deliberating and explaining herself, we would find that this person led an extremely impoverished life, and we would be inclined to declare the person insane. All she could do was act on her strongest desire at any given moment. Such a person could never articulate a genuine reason for her actions; all she could say was that she did something because her strongest desire made her do so. In that sense, she could only refer to causes and not to reasons. If acting means acting for a reason, such an individual could not *act* at all. She could not be said to have a fully developed human mind.

As persons, we not only have desires, but also desires about which desires to have (second-order desires) (Frankfurt, 1988). Perhaps I find that every time I meet the Salvation Army, my miserliness makes me walk by without donating anything, but when I come home I regret it and decide that the next time I meet them, I would like to have the desire to give away some money. So we have the capacity for evaluating our desires. But in light of what? If we could only evaluate desires in the light of other desires, as in the restaurant example above, it would merely be a matter of determining which desires were the strongest. Then we could never have a *reason* to change our desires, and this runs counter to our everyday moral experience where we in fact are concerned with changing our desires for moral reasons, and not just because other desires are quantitatively stronger. Taylor here introduces his notion of strong evaluation. Strong evaluation is "when the goods putatively identified are not seen as constituted as good by the fact that we desire them, but rather are seen as normative for desire" (Taylor, 1981:193). In strong evaluation we are concerned with the qualitative *worth* of our motivations, desires, and ways of life. The "strength" of our desires does not matter here, but rather the issue of whether what we desire is worth desiring, whether it is *desirable*.

The correctness of strong evaluations does not depend on the subject's attitude (Laitinen, 2002b), but on the moral values and moral reasons for action that there are *independently* of the subject. We identify moral reasons, not by deciding which preference to realize, but by understanding what Taylor calls "the import of a given situation" (Taylor, 1985b), that is, we perceive what in the situation gives the normative grounds for our feelings and actions. These "imports" (or moral properties) can only be described adequately in moral terms by way of thick ethical concepts.[13]

[13] I do not think that it makes much sense to say that there are moral values and reasons wholly independently of human as such. This is contrary to the moral realism of Robinson (2002), which is not a version of "relational realism," but resembles Platonism. I believe that moral properties are real, but they require embodied, experiencing, and engaged human beings, just as smiles require lips and mouths, yet are no less real for that. But once there are humans able to act, that is, able to articulate the reasons confronting them in specific situations, then there are moral properties in the world. In this sense, my perspective can be characterized as a *mutualist* perspective, operating with a mutuality of the agent and the environment (in a broad sense including social and moral affordances) (cf. Still & Good, 1998).

Taylor (1989:59) makes two points in this connection, the second of which I have already mentioned: "(1) You cannot help having recourse to these strongly valued goods for the purposes of life" and "(2) What is real is what you have to deal with, what won't go away just because it doesn't fit with your prejudices." What he means by the first point is that strongly valued goods are a transcendental condition for the existence of acting human beings as we know them. If we could not refer to properties of the world that are morally valuable independently of our desires and motivations, then we could not conceive of human action. This is so because identifying human action implies identifying reasons, and these in turn refer to moral properties, that is, issues of value and worth that stand or fall independently of my personal inclinations.

The other point made by Taylor is that what is real is what we have to deal with. We have to deal with morality in our lives, and we need the kind of understanding of the world that can only be expressed in moral concepts: "What better measure of reality do we have in human affairs than those terms which on critical reflection and after correction of the errors we can detect make the best sense of our lives?" (Taylor, 1989:57). And further:

> Suppose I can convince myself that I can explain people's behavior as an observer without using a term like "dignity." What does this prove if I can't do without it as a term in my deliberations about what to do, how to behave, how to treat people, my questions about whom I admire, with whom I feel affinity, and the like? (Taylor, 1989:57).

To sum up: My point has been that value terms pick out real features of the world (moral features) and that we could not understand ourselves, other people, or human action in general without them. They are indispensable to us, and we should therefore not try to remove them or purify them in psychological studies of human life, for they are an essential part of the language we need to act and live.

Moral Re-Action – The Moral Normativity of the Emotions

With the argument rehearsed above, we can conclude that moral values are presupposed not just in human *action*, but equally so in human *emotion*, for, like actions, emotions are normative. They are "intelligent and discriminating parts of the personality, closely related to beliefs of a certain sort, and therefore responsive to cognitive modification" (Nussbaum, 1994:78). As Aristotle famously said:

> It is possible, for example, to feel fear, confidence, desire, anger, pity, and pleasure and pain generally, too much or too little; and both of these are wrong. But to have these feelings at the right times on the right grounds towards the right people for the right motive and in the right way is to feel them to an intermediate, that is to the best, degree; and this is the mark of virtue (Aristotle, 1976:101).

The excellent or virtuous human being is properly affected. It is proper to fear death, for example, for death is worthy of being feared, but a phobic fear of pigeons is normatively wrong. Harré (1983:221) has noted that several of the seven deadly sins are emotions: sloth and gluttony are indeed actions, but lust, anger, envy and pride are emotions. To view emotions as normative is complementary to the account of actions given above. The human world is a normative order, and emotions are a primary way of being tied to that order (Baerveldt & Voestermans, 2005). In our everyday lives, as De Sousa (2001) has argued, emotions are both subject to moral evaluation *and* an aid in moral evaluation and cognition.[14]

This view of the emotions go against the current orthodoxy in psychology and the cognitive neurosciences that would have us focus on the neurological and physiological aspects of emotions (e.g., Damasio, 2003).[15] This orthodoxy overlooks, however, that many emotions do not involve changed physiological conditions at all (Hacker, 2004). I can fear that the inflation rate will rise in my country, for example, without this fear manifesting itself in any bodily perturbations, and I can feel proud or lonely without any corresponding physiological processes (Brinkmann, 2006a). Needless to say, many emotions come with a distinct physiology, but it seems highly questionable that we can identify emotions through introspection or by being aware of bodily changes alone. Rather, we identify emotions by recognizing and assessing their *objects* (we fear *something*, take pride in *something*, and are angry with *someone*) – and these exist only within some normative order.

In this light, emotions have an epistemic dimension. They involve cognition, not of *what* the world contains (e.g., loved ones and dogs), but of *how* the world is (e.g., depressing and dangerous). From the normative point of view, emotions must be seen as more or less adequate responses to the events of our social worlds. Emotions embody the moral values, thoughts, and judgments of a culture:

> Most emotions reflect – to a greater or lesser extent – the thought of an epoch, the secret of a civilization. It follows that to understand the meaning of an emotion is to understand the relevant aspects of the sociocultural systems of which the emotion is a part (Averill, 1982:24).

Conversely, moral values and ideals can only be upheld in a culture if they resonate emotionally in individuals. Emotions, ideals, and the moral order are deeply interrelated. Guilt is perhaps the central emotion that connects us to the moral order, for it is the emotion a person feels when she has the experience of being the source of some wrongdoing, relative to a local moral order. Guilt need not ensue in such situations (there is no direct, mechanical link involved), but there is at least a normative reason to feel guilt. Other emotions can also accompany the experience of having acted wrongly. One can be angry with oneself, be embarrassed, or shameful. But guilt is primary, for if one becomes angry with oneself in such a situation, it is *because* one did something that elicited guilt, and not the other way around (unless, of course, it was the very display of anger that constituted a breach in the shared

---

[14] It is important to distinguish between the view that the fact that I feel positively towards something *constitutes* the positive value of this something, and the view that emotions can *disclose* certain value-laden facts of the world for us. The former view is emotivism, whereas the latter view goes back to Aristotle. The former view says that emotions always *define* value, the latter that emotions can *point* to value.

[15] The following builds on Brinkmann (2009).

norms). It is the normativity of guilt that explains why we normally praise those who actually do feel guilty, when they have acted wrongly. That is, we praise such people because they are able to feel guilt (and not, of course, because they have acted wrongly). And we condemn those who have acted wrongly, but do not feel or express any sense of guilt. If guilt is connected to moral transgressions, we can approach this emotion as perhaps the most significant probe into the moral experiences that are prevalent in a culture.

Positioning theorists such as Gerrod Parrott (2003) have analyzed the links between social positioning (the ongoing negotiations of rights and duties that occur in our moral orders) and the emotions.[16] Parrott describes how emotions can play different roles in positioning. One way of positioning *oneself* "is to display the emotions that are characteristic of one's position" (p. 29). And a way of positioning one's *opponents* "is to state what emotions they ought to be feeling and to characterize as inappropriate the emotions they are feeling" (p. 29). Thus, ascribing emotions to oneself and to others is a normative process and a central aspect of social life. It is also a significant part of human socialization, where parents and educators consistently attempt to position children by stating which specific emotions they should appropriately feel. We say such things as "Aren't you happy?," "You should be ashamed of yourself!," and "There is nothing to be afraid of!." If emotions were pure causally induced states in organisms that were suffered in a passive, mechanical way, such forms of normative emotional positioning would be unintelligible.

The complex character of the normativity of emotions can be seen from considering some of the positioning analytic distinctions. One distinction can be drawn between "counter-emotions" and "relabeled emotions." Counter-emotions are emotional expressions that contradict a positioning that was inherent in a preceding emotional expression (Gerrod Parrott, 2003:33). If A is angry at B, for example, then B is positioned as blameworthy, for anger is legitimately felt in response to experienced wrongdoings. Thus, according to the conventional normative choreography of emotional episodes, B ought in such a case to express the emotion of guilt and perhaps offer an apology. B is here positioned as guilty, but B may also react by re-positioning A through expressing a counter-emotion, for example anger. This would position A as the wrongdoer in the sense of having felt *unwarranted* anger. An alternative option for B would be to accept responsibility for the action that made A angry, but at the same time maintain that the action was justified, and perhaps express pride rather than guilt. Relabeled emotions are often found in political contexts, when the anger expressed by one group is relabeled as envy by another group ("the protesters have no right to be angry – they must really be envious of us"). We can see the same (paternalistic) process of relabeling occurring

---

[16] Positioning theory is a recent variety of discursive psychology, which was originally developed as a dynamic alternative to theories that stress roles and rules (Davies & Harré, 1999). Rather than invoking static roles that determine how individuals act, positioning theory argues that the concept of positioning can bring forth the dynamic and changeable aspects of social life, where discourses are seen as offering subject positions that individuals may take up – or may be forced into.

when parents are positioning misbehaving children as tired. Relabeling the emotions of others is a very powerful way of positioning them, for example as immature or uncontrollable.

We see that emotions, positionings, and moral orders are deeply related, and that ongoing negotiations and struggles over the normativities of social life centrally involve emotions. Disagreements about emotions are first and foremost normative disagreements, that is, about the legitimacy and correctness of how we feel, and *ought* to feel, in relation to certain episodes, their values, and reasons for action.

Moral Identity

We shall now return to the question of moral realism. So far, the argument in favor of moral realism has entailed the rejection of two common forms of reductionism: One that wants to reduce our moral reactions to brute, instinctual reactions (the evolutionary view), and one that wants to find descriptive equivalents to our moral concepts and understandings (a scientific fact-value dichotomy). I will now, rather more briefly, consider a third and related kind of reductionism. This is the view that moral values are ultimately the result of choice. The discussion of this position will lead me to the argument that human identity is irreducibly structured by moral values; values that cannot be construed as something of the agent's own choosing. We are "thrown" into moral reality as Heidegger could have put it. Morality was there before me; morality is something for me to discover and become adequately connected to, and the quest to do so is a quest for identity.

In his well-known paper "What is human agency?" Charles Taylor (1985d) discusses Sartre's notion of radical choice. Radical choice is a choice that is not grounded in any reasons, but simply willed. Sartre believed that our values should be thought of this way: as the result of radical choice. In *L'Existentialisme est un Humanisme*, Sartre tells the story of the young man who faces the dilemma of remaining with his ailing mother or going away to join the Free French in Britain. Sartre is unable to offer advice to the young man, because in his view the issue has to be settled by radical choice. There can be offered no genuine reasons in favor of either option, and the young man should therefore just choose one option in the face of this existential absurdity. What Sartre forgets, as Taylor notices, is that this situation is only a dilemma for the young man because he is faced with two rival moral demands. His mother might die if he leaves her and at the same time the enemy is about to destroy the foundation of ethical life in the young man's country, and he cannot just sit passively and watch that; "it is a dilemma only because the claims themselves are not created by radical choice," says Taylor (p. 30).

If the values that give rise to the dilemma were in fact the result of a choice, then the young man could at any moment dissolve the dilemma by declaring one of the claims invalid. The fact that he cannot do so brings out the ludicrousness of the idea that values are created by our choices; "if serious moral claims were created by radical choice, the young man could have a grievous dilemma about whether to go and get an ice cream cone" (Taylor, 1985d:30). What is valuable and what is not, is not determined by our decisions. These issues "resist" our dealings with them and "object" to us. Even Nietzsche's nihilistic remaking of the table of values, his attempt to *redefine* what is of worth, presupposes a horizon of moral issues that he did not himself invent or choose. Remaking the table of values "means redefining values concerning *important questions*, not redesigning the menu at McDonald's, or next years casual fashion" (Taylor, 1991:40; my emphases, SB).

This is a powerful argument against voluntarism in general, that is, the view that values result from our choices and not just against Sartre's existentialist ethics. The argument reiterates the earlier point that we cannot do without some sense of moral orientation or strong value (Smith, 2002:93). As one of Taylor's exegetes says, "no one could conceivably choose to live a 'care-free' life in the sense of a life led without background distinctions of worth" (p. 100). We cannot as humans do without an orientation to un-created and un-chosen values. But what would we lack without such an orientation? We would lack an identity, Taylor responds. How is that?

When we are asked to describe who we are, we are supposed to describe what is constitutive of our identities, but how to answer it is not something that we can individually decide. If I am asked to describe who I am, and then say that I am someone with 684,923 hairs on my head, this will not be taken as a satisfactory answer even if it is correct. What we look for, in response to such a question, is an articulation of what is important to me. A sensible answer would thus state that I am a husband, a father, a teacher, a Danish citizen, someone who is trying to write a book on psychology as a moral science, etc. "To know who I am is a species of knowing where I stand," says Taylor (1989:27). And "where I stand" is inevitably in relation to issues having to do with values:

> My identity is defined by the commitments and identifications which provide the frame or horizon within which I can try to determine from case to case what is good, or valuable, or what ought to be done, or what I endorse or oppose (Taylor, 1989:27).

My real-world identity is bound up, not with an inner choosing will or a private self, but with the moral framework within which I find myself. "Our sense of who we are is linked to the stand we take on issues of concern, and for that we need points of orientation, the reference points provided by frameworks of qualitative contrast" (Smith, 2002:97). Such frameworks are thus *given* us to discover and articulate rather than something to choose. They are based on traditions and communities, on shared practices and institutions, and not on choice or self-reflection.

If our identities are formed by our commitments to issues of moral worth, this does not mean that identities are fixed once and for all. On the contrary, having an identity, living a life, should be seen as a quest (MacIntyre, 1985a), as a kind of craving to be rightly placed in relation to the good (Taylor, 1989:44). In order to understand this temporality, we need narratives (Bruner, 1999; Laitinen, 2002a; Ricoeur, 1992) contextualization, and thick description (Levine, 1998). Our identities are framed by our relations to values, and the crux of our biographies is how we move and develop in relation to these values: "It only makes sense to ascribe direction to a life if we can distinguish between more or less significant moments, events, or experiences. But in doing this we are articulating a changing relation to the good" (Smith, 2002:98). Narrative identity is what makes sense of our movements in the space of moral reasons.

Morality as a Social Construction?

There is, however, a final challenge to the interpretive–pragmatic moral realist view. This challenge comes from social constructionism, and I shall here concentrate on Kenneth Gergen's well-known and well-articulated version (1992, 1994, 1999, 2001a, 2001b). The core argument for social constructionists is that meaning is a product of social discourse, so we can never come in contact with a world that transcends our social constructions: "Theoretical accounts of the world are not mirror reflections of the world but discursive actions within a community" (Gergen, 2001a:811). Consequently, moral phenomena, concepts, and actions are also socially constructed in ongoing interaction and negotiation, and moral statements are seen as poetical and rhetorical devices, "among the resources available for playing the games and participating in the dances of cultural life. They are plays or positionings that enable persons to construct the culture in what we take to be a moral or ethical way" (Gergen, 1992:17). As Vivien Burr says, echoing Gergen: "the agency of human beings lies in their ability to manipulate discourse and use it to their own ends" (Burr, 1995:92).

It was argued in Part I of this book that psychological categories refer to human kinds, rather than natural kinds. This is essentially a social constructionist idea, and the question now is why a similar line of analysis should not be applied to moral phenomena? Why do I want to uphold a realist position as regards moral phenomena, when I have insisted that psychology makes up people by being engaged in the game of looping effects of human kinds? Is it not true that just as psychological phenomena come into being along with their relevant "fields of description," so it is with moral phenomena? Are not these arbitrarily constructed in the course of human interaction as it unfolds in different cultures? My answer to this is no, and I shall end this chapter on moral realism by spelling out why I believe that we have good reasons to reject the suggestion that morality is a social construction.

The Wittgensteinian Legacy

Social constructionists put emphasis on the *linguistic* construction of our reality. They often seek philosophical legitimatization for this view in Wittgenstein's (1953) later works. In Wittgenstein's first book, the *Tractatus*, he sought to provide a general theory of linguistic meaning by arguing that sentences are linguistic *pictures* of reality that represent the world in virtue of their logical form. Semantic necessity was thus construed as *logical* necessity, and the essence of language was sought in its capacity for logically picturing the world (Holiday, 1988:31). In his later philosophy, as crystallized in the *Philosophical Investigations*, the project of finding the essence of language was given up. Language has no essence, Wittgenstein now argued, for it is used in all sorts of ways, in all sorts of practices ("language games") that have widely different rationales. Linguistic meaning is not to be sought in the way language *represents* the world, but in the way language is *used* in social practices. Understanding what language is should not involve the search for an answer to how it succeeds in making contact with "the real world," for in Wittgenstein's later philosophy "to imagine a language means to imagine a form of life" (Wittgenstein, 1953:§19). Language gets meaning, point and purpose from belonging to many different and diverse human practices. Semantic necessity is no longer an essential logical necessity, but something much more heterogeneous.

Anthony Holiday has argued that semantic necessity in the later Wittgenstein must be seen as *moral necessity*: Language, as we know and speak it, depends on a number of regularities; some are *natural* and some are *normative*. Natural regularities are those we find in the physical world. We can only speak coherently about objects and events, because these behave in fairly regulated ways. If there were no natural regularities, if, for example, the weight of things changed constantly, then we could not talk about "the weight" of anything, and all our weighing practices would be meaningless. Clearly, the material aspects of the world set limits to the conceptual possibilities that govern language (Holiday, 1988:67). But Holiday is more interested in the *normative* regularities that he finds also govern language (and, I would add, psychological phenomena). He argues that certain language games, which he calls "core-language-games" are essential to any language imaginable. Core-language-games function to preserve certain moral values without which language would lose its communicative force, and hence its meaning. He identifies three such core-language-games:

1. *Truth-telling language games*: In general, we praise truth-telling and condemn and punish lying. We have linguistic practices that function to preserve the value of truthfulness, which, in this light, is not a value that can intelligibly be seen as socially constructed, but rather *presupposed* by any process of social construction. For there to be social constructions, there must be a language, and language is only imaginable if people are committed to truth-telling. It is a fundamental fact, as Løgstrup (1956) also argued, that a basic trust is primary in social interactions and conversations. This basic trust may of course be subdued from time to time, but it is nonetheless ontologically primary. Humans expect each other to tell the truth, for lying is logically parasitic on truth. If humans normally lied, there could be no such thing as language. Even in a world that is affected by the representations (social constructions) humans make of it, there is still room for an objective morality. In Hannah Arendt's words: "even if there is no truth, man can be truthful, and even if there is no reliable certainty, man can be reliable" (Arendt, 1958:279). The constructionist insistence that meaning is a convention that is endlessly negotiable, and that everything that is said could be otherwise (Gergen, 1999), does contradict the basic fact that humans have to be committed to truthfulness and sincerity in their concrete dealings with each other if language is to be possible (I expand on this critique in Brinkmann, 2006c). As Holiday

says, paraphrasing Peter Winch, adherence to the truth-telling norm "is not itself conventional, but the condition of there being any conventions whatsoever" (Holiday, 1988:93).

2. *Justice language games*: Justice language games are those we engage in to determine guilt and innocence in non-arbitrary ways. The practices of courts of law are institutionalized versions of such language games. Holiday's point is that if Wittgenstein was correct to insist that to imagine a language means to imagine a form of life, then justice, like truthfulness, is a *pre-conventional* value, and *not* a socially constructed value. As a value, it is a *condition for* our having a language, rather than something constructed *in* speaking. If there were no objective distinctions between guilt and innocence, "it would not be possible to distinguish harm-attracting activities from safe ones" (Holiday, 1988:105). There would thus be no reason to suppose "that such-and-such a language game was a harmless one to play, or that this or that way of trying to act in accordance with a rule, though it did us no harm yesterday, would not be treated as treason tomorrow" (p. 105). We would thus not be able to engage in language games at all, and language would be impossible.
3. *Ritual language games*: If there is to be language, Holiday argues, there has to be "deep-seated agreement which is only possible if the integrity of the persons who speak the language is sustained, and clearly this cannot be done unless reverence for persons and their rights to speak and be listened to is a prevailing norm" (Holiday, 1988:109). Such reverence presupposes a respect for ritual. Holiday refers to anthropologist Mary Douglas's view of rituals as enactments of social relations enabling people to know their own society. In this view, rituals are necessary to create and maintain a particular form of life, and insofar as language belongs to a form of life, the non-conventional value of respect for ritual is presupposed in our having a language in the first place. Like truthfulness and justice, such respect is a *transcendental* condition for the possibility of human language.

In short, Holiday's argument is a powerful rejoinder to the view that moral ideas are socially constructed. Instead, we have to admit that there are objective moral values that make language possible. Core-language-games are those that serve to preserve and sustain these moral values and also the complex psychological life made possible by them.[17] The utterances of moral skeptics and constructionists "are undeniably framed in the medium of language, and that medium has been shown to be unimaginable without, and to depend for its coherence upon, certain value-laden

[17] The core-language-games serve to protect an elementary moral normativity in human life. The Danish theologian and philosopher, Løgstrup (1956) referred to this elementary normativity as "the ethical demand." This naturalistic demand springs from the fact that human lives are interdependent; we inevitably have some power over each other, and we can use this power to serve our neighbor – or the opposite. Løgstrup states that the ethical demand is a pre-conventional and pre-cultural demand to serve the other, which is prior to (thus not created by) our feelings, thoughts, or will. I believe that Løgstrup's phenomenology of the ethical demand is quite similar to the analysis of moral realism in this chapter. The next chapter seven is not about the elementary normativity of the ethical demand, but about what Løgstrup referred to as *morality*, moral normativity as instantiated in social practices.

practices and moral certitudes" (Holiday, 1988:111). In my own words, we not only live in a moral world because we have language and a complex mental life, but we have language and a complex mental life because of objectively real moral values. The space of reasons is constituted by these moral values.

Conclusions

I can now conclude on the realist view that has been defended in this chapter, and sum up its implications for the interpretive–pragmatic outlook on psychology and morality. It has been argued that moral values are among the objectively real things that structure mental life.[18] We cannot make sense of human perception, action, emotion, and identity without presupposing that moral values are real. Morality is part of the fabric of mental life – even for those who deny it. I hope to have provided at least a gist of how to think about these psychological notions in a moral realist context – which is the project of the proposed interpretive–pragmatic psychology. I have also argued that morality is irreducible. It can neither be accounted for in terms of survival and fitness (sociobiology) nor be accounted for in terms of contingent social constructions (social constructionism). We cannot take moral phenomena out of the space of reasons and still expect them to be *moral* phenomena. As it happens, morality is presupposed in every attempt at denying its reality.

The form of interpretive–pragmatic moral realism defended here is far from a transcendent Platonic realism that insists on an eternal and unchangeable idea of Good, but nevertheless it is realism since it operates with the notion that there are objective moral values that determine what counts as good reasons for action and emotion. Morality is real in the sense that we have to deal with it in living our lives (this is the pragmatic side of the proposed moral realism), but it is not self-evident like logic, but exists in historically evolved social practices (this is the hermeneutic or interpretive side[19]). The specific role of social practices in interpretive–pragmatic moral psychology is taken up in Chap. 7.

---

[18] I say "among" here because I certainly do not want to deny that other things in addition to moral values are necessary in order for mental life to exist, for example, language, community, bodies, and brains!

[19] Laitinen (2002b) has perceptively referred to Taylor's hermeneutic moral realism as "culturalist moral realism." It could also be called "relational moral realism."

Chapter 7
Moral Practices

Introduction

In the previous chapters, it has been argued that psychologists ought to think about morality not only as something real, but also as something that co-constitutes psychological phenomena. Now, the question is: In what way? Can we specify the mode of existence of morality and moral phenomena? I will attempt such specification in the present chapter by following up on the previous two chapters and argue that practices are the sources of moral intelligibility. This is contrary to two other dominant views according to which emotions and desires (cf. Hume) and universal rational procedures (cf. Kant) are the sources of moral intelligibility. By linking morality to practices we are able to think of it as real and objective and at the same time as something tradition bound with a cultural history. We can perhaps think of morality as something relative without being relativists (LaFollette, 2000).

In the last few decades, the social sciences have been undergoing a practice turn. This is found in sociology (e.g., Giddens), anthropology (e.g., Bourdieu), science studies (e.g., Hacking, Latour) and philosophy (e.g., the contemporary followers of Dewey, Heidegger and Wittgenstein). There is in practice theory an anti-dualistic thrust against any hard-and-fast distinctions between organism and environment, subject and object, actor and structure, individual and community. These dualisms are dissolved through the notion of practice: Organism and environment are understood as abstractions from a more fundamental practical transaction in which these entities are inseparable (Dewey & Bentley, 1949). Actors and structures are not diametrically opposed, and neither has existence outside those social practices that shape them. Individuals and communities are not separate essences, but emerge from a more fundamental plane of practices.

If it is possible to condense one common idea behind the plethora of different practice theories, it must be the idea that humans meet the world in active engagement. We do not passively absorb the structure of the world, neither "filtered" (as in rationalist conceptions of knowledge) nor "unfiltered" (as in empiricist conceptions of knowledge). Knowledge is not a kind of *representation* of the world at all, but a kind of *doing*, and human doings are not discrete events, performed by isolated actors, but intelligible only as parts of larger wholes: practices. Practices can thus initially be

S. Brinkmann, *Psychology as a Moral Science: Perspectives on Normativity*, DOI 10.1007/978-1-4419-7067-1_7, understood as coordinated human activities, in some community of human agents, involving a certain temporal extension. Psychology has by and large remained outside the practice turn. With a few exceptions that I will return to later in this chapter, this also includes psychological theorizing about values and morality. The interpretive-pragmatic view developed in this book can, however, make use of the valid insights of practice theory in relation to moral and psychological phenomena.

In order to characterize the interpretive-pragmatic view of morality that I wish to espouse, I will begin by briefly contrasting it with two other models of reasoning in general, and of moral reasoning in particular. These I call instrumentalism and proceduralism (roughly Humean and Kantian). These depict moral reasoning as means-end reasoning and procedural reasoning respectively. I argue that both are defective and that a practice approach can remedy their problems. Then I concentrate on the notion of practice, which can be given either a descriptive or a normative interpretation (as either "regularity" or "normativity"). I present and defend a normative understanding of practices, which goes back to Aristotle and has been revitalized in recent years, particularly by Alasdair MacIntyre's theory of practice.

Three Conceptions of Moral Rationality

Western thought has produced three broad conceptions of reason and rationality (Taylor, 1975):

1. A substantive view that goes back at least to Plato, where reason is understood "as the power by which we see the true structure of things, the world of the Ideas" (Taylor, 1975:367). Reality is structured in a definite way, and our rational powers disclose the structure of reality for us. Practical rationality, which is that part of our reasoning that is concerned with acting, helps us act according to the pregiven structure of reality. A variety of the substantive view is Aristotle's. Like Plato, Aristotle held that being practically rational means acting on the basis of certain substantive values, but, unlike Plato, Aristotle argued that these are more varied than a single Platonic Idea of Good. The practice-oriented approach to morality (to be presented below) follows Aristotle on this point and argues that substantive moral values are inherent in social practices, and *not* in a transcendent ideational realm.
2. The other view is the instrumental view, where it is claimed that "practical reason is the intelligent calculation of how to encompass ends which are beyond the arbitration of reason" (Taylor, 1975:367). This means that reason ultimately cannot help us in moral decision-making, for, according to this perspective, ends are outside the scope of reason. Hume's subjectivism, Mill's utilitarianism, and the emotivist theories of the twentieth century are varieties of this perspective, which is an important part of the psychological social imaginary, as we have seen in part I of this book. There is no longer a normative order of things in the world to serve as the basis for moral judgment. We choose our ends – or merely *have* them as brute natural facts – and rationality is solely concerned with finding the

optimal means for their realization. Thus, there is a clear separation of means and ends, and reason is operative only concerning means. This view has historically influenced scientific psychology through Hume and Mill. According to Gazzaniga and Heatherton (2003:18), Mill is responsible for the fact that psychology left the realm of speculation and became a science of observation and measurement. Many sociobiologists and evolutionary psychologists tacitly build upon the Humean approach to practical reason, but one does not have to be a sociobiologically inclined psychologist in order to subscribe to Hume's "sentimentalism" (Robinson, 2002). A few examples may suffice here: An influential researcher like Martin Hoffman, who is well-known for his studies of children's development of empathy and morality, claims that moral principles are validated by feelings. Feelings and empathy are what motivate us to act morally. He rightly refers to Hume as the predecessor of this view (Hoffman, 2000). Likewise, Nancy Eisenberg, whose field is prosocial behavior and altruism, refers to Hume as a main background figure (Eisenberg, 1996:52). In line with Hoffman, she gives feelings and sentiments the leading role in moral psychology (see Eisenberg, 2000).

3. The third view is the procedural view, which differs from the instrumental view in claiming that rational procedures in fact dictate not only means but also ends, so ends can also be rationally evaluated. But in light of what? Not in light of substantive values, for Kant, who incarnates this third procedural view, eschews substantive values as much as the proponents of the instrumental view. Kant "does not accept any sort of substantive value realism. [...] We do not want things because we perceive that they are good: rather our initial attractions to them are natural psychological impulses" as Korsgaard, an influential interpreter of Kant, says (1998:52). With Kant, no natural goods or substantive normative orders are accepted. Something counts as good only if based on a purely formal criterion: does it bind a rational will? Can I rationally will that what guides my action should count as a universal law of nature? If so, what I do is good. This rational procedure, also known as the categorical imperative, is therefore that in light of which we can rationally evaluate our ends. In moral psychology, Kohlberg and Turiel have been two significant researchers within the procedural tradition. Before proceeding to the practice view of morality, let me briefly sketch the ways in which the procedural view has influenced Kohlberg's and Turiel's psychological inquiries into moral matters.

Procedural Moral Psychology

Kohlberg's study of moral development (Kohlberg, 1981, 1984) is still the paradigmatic example of a psychological treatment of the moral field. He worked on the basis of Piaget's constructivism, which also owed much to Kant (Piaget, 1932). According to Piaget, morality is a system of rules, and the task of genetic (i.e., developmental) moral psychology is to determine how we as individuals come to internalize this system of rules, and learn to respect them. Kohlberg constructed a stage theory of moral development beginning with *preconventional morality* (stages 1 and 2). This stage is characterized by what Piaget called "moral realism": the alleged childish view that moral qualities are inherent in actions just like physical qualities (e.g., size) are inherent in objects. In this immature stage, justice is conceived as *quid-pro-quo*, give-in-order-to-get, often masking underlying egoistic motives. The next stage is *conventional morality* (stages 3 and 4), where a more reflective understanding is developed, and the child becomes able to see the world from other people's point of view. Moral rules are now seen as ways of maintaining social order. Moral development culminates with *principle-oriented morality* (stage 5 and a hypothetical stage 6), which is a form of deontological universalism *pace* Kant. The rational moral agent is here capable of identifying phenomena like freedom and dignity as universalizable values, completely independent of, and prior to, the actor's participation in concrete societal and cultural processes. It is thus a society-*creating* perspective (Lapsley, 1996:74). The autonomous actor is seen as capable of deducing a just societal order from the procedures of his or her universal reason alone.

Kohlberg outlined a number of meta-ethical assumptions as the philosophical foundation of his empirical research (Kohlberg, 1984:215–216). He argued for example that moral development is universal across cultures and that moral statements can express knowledge and cannot be reduced to emotional exclamations (as in emotivism) or de facto preferences (as in Hume's sentimentalism). Contrary to emotivist or sentimentalist theories, emotions are seen as irrelevant when we consider moral action. What render moral statements moral are formal criteria. It is the procedure involved, rather than the substantive content of our judgments, which determine their moral qualities. Moral reasoning builds on general rules and principles, according to Kohlberg. He rejected what he called "contextual relativism" (p. 296) and he understood moral principles as constructions arising from the universal and rational structures of the human mind. This must be distinguished from the kind of *social* constructionism that was discussed in the previous chapter, for social constructionists are suspicious of the idea of *individual* minds constructing anything in independence from social and cultural processes. Finally, the function of moral principles is to solve social or interpersonal conflicts in Kohlberg's eyes. Morality is thus not about *the good* but exclusively about justice, *the right*. As he said:

> I make no direct claims about the ultimate aims of people, about the good life, or about other problems that a teleological theory must handle. These are problems beyond the scope of the sphere of morality or moral principles, which I define as principles of choice for resolving conflicts of obligation (Kohlberg, 1981:169).

Kohlberg's theory respects the fundamental normativity of morality. However, his theory has been criticized on a number of other points, and I shall very briefly present some of them here. First of all, it has been argued that his theory lacks psychological realism (Flanagan, 1991). It does not give an accurate picture of what we actually do when we reason morally. Gilligan has demonstrated, in her now classical study (Gilligan, 1982), that we are not normally hyper-reflexive moral reasoners who follow strict and impartial moral principles. In everyday moral living (contrary to Kohlberg's artificial "Heinz-dilemma"[1]), morality is not just about abstract justice, but also about care and interpersonal relations (all those things about "the good life" that Kohlberg confers to "a teleological theory" that he does not want to touch). Perhaps it is all for the best that we are not such hyper-reflexive moral reasoners, for, as Williams (1985) has argued, it would ruin our chances of leading good and satisfying lives if the abstract, impersonal Kantian reason should decide every particular action. It would in fact make self-alienation a central moral virtue, namely in the form of alienation from one's own concrete relations and commitments.

In addition, Kohlberg has not accounted for the relation between moral reasoning and moral action. It has been shown that even very good and altruistic human beings do not necessarily score well on Kohlberg's scale (Colby & Damon, 1992). Furthermore, Kolhberg's theory, which was conceived as universal, has had major problems in understanding moralities of non-Western cultures. Buddhist ethics, for example, are radically different from both Kohlberg's ideas (Huebner & Garrod, 1991).

The main problem, however, is the general one inherent in all procedural moral systems: How to connect the abstract rules and principles (generated by the moral procedure) to concrete social situations? A *locus classicus* of this problem (sometimes called the *subsumption* problem) is Plato's *The Republic*, where Socrates demonstrates that even if we succeed in formulating general moral principles like "one should always return borrowed items to their owner," we still have to apply such principles in concrete situations. No principles are self-interpreting or self-applying. In a situation where the item is a weapon, and the owner is out of his mind, this principle does no longer apply (Plato, 1987:65–66). And so it is with any moral rule conceivable. Rules need to be applied to concrete situations, but we cannot forever formulate rules for how to apply rules, because at some point we have to *act*. Kohlberg's proceduralism does not adequately tackle this issue of moral application.

In fact, as Dreyfus and Dreyfus (1990) have tried to demonstrate, a morally proficient reasoner (the Aristotelian *phronimos*) rarely engages in reflective rule-following, but sees clearly and intuitively what to do in concrete situations (cf. the earlier discussion of moral perception). It seems to be the case that a "thick" understanding of concrete social practices is necessary to understand and judge morally. We need to know when and how to employ our principles and procedures, and this kind of knowledge cannot itself be procedural, but must be based on a background understanding of substantive values. If so, certain substantive values are in fact presupposed by procedural theories. In Kohlberg's case, it is quite clear that justice plays a key role, but what should we do when this (substantive) value comes into conflict with other values, like equality, general welfare, or concrete bonds to other people? There are good reasons to think that we need "thick" and elaborated teleological understandings of the goodness of

[1] Where subjects are presented with the hypothetical dilemma of choosing between two evils: stealing a drug or risking the death of a loved one.

human life in order to act responsibly in such situations. A "single-term morality" of this procedural kind will not suffice (Taylor, 1996).

Elliot Turiel's theory, which has been developed in collaboration with a number of colleagues (Helwig, Tisak, & Turiel, 1990; Helwig & Turiel, 2002; Helwig, Turiel, & Nucci, 1996; Nucci & Turiel, 1978; Turiel, 1983), is perhaps not as well-known as Kohlberg's, but it is equally interesting. Like Kohlberg, Turiel is inspired by Piaget and Kant, but he uses them somewhat differently. One can say that Turiel puts Kohlberg theory "on the side" (Shweder, Mahapatra, & Miller, 1990), for rather than seeing three levels of moral understanding that follow each other in the course of development (egoism, conventional morality, and genuine morality), Turiel finds that these all exist side by side in the course of ontogenesis. He interprets these levels as *knowledge domains* (Turiel, 1983), and refers to them as the *personal domain* (e.g., concerning which color of shirt to wear, based on personal preferences), the *conventional domain* (e.g., concerning whether to wear a bathrobe to school, based on social coordination in social systems) and the *moral domain*, which, following Kant, is defined as encompassing rights, justice and welfare. Contrary to Kohlberg, Turiel has found that children can distinguish between these domains of knowledge at a very early age; perhaps from the age of two (Helwig et al., 1990). Turiel argues that this is due to the fact that children have quite different forms of experience with interpersonal interaction in these different domains (Turiel, 1998). Children learn quite early that morality does *not* depend on social conventions, for they are punished for behaving immorally, also in situations where no social conventions are known or explicitly formulated. Like Kohlberg, Turiel believes that his distinction (between the three knowledge domains) is universal across history and cultures. Morality simply *never* has to do with specific cultural practices.

Cultural psychologists have criticized Turiel heavily on this point (Shweder et al., 1990). According to Shweder et al., it is impossible to distinguish in any universal way between convention and morality. Not that they are the same (Shweder, 1990, defends a form of moral realism), but rather that the specific *ways* of distinguishing vary across cultures. Not all cultures agree with the West in the first place that social practices are conventions. Shweder finds that only those cultures that focus on the autonomous individual self as a carrier of personal rights *prior* to concrete participation in practices agree that practices are conventions. Shweder argues that the Western idea that practices are conventional is connected to the dominance of the market in our culture. The market as an institution is wholly neutral as regards specific social arrangements that individuals choose to engage in (p. 132). In the West, the typical idea (our social imaginary) states that all social arrangements are derived from a more fundamental authority: the universal reason and rights of the individual. We typically conceive of social relations as instrumental for individual preference satisfaction, but this is very different in other parts of the world. Thus, Turiel's conception of three entirely different knowledge domains is perhaps a Western prejudice.

In the Hindu culture that Shweder has studied extensively, the worst moral transgression (among 39 possible ones) is if the eldest son has his hair cut and eats chicken the day after his father's death (Shweder et al., 1990:165). Turiel would object that this can never be a *moral* judgment, but merely something based on a convention related to the Hindu worldview and its social practices. But the same line of response could be used against Turiel: Something is a moral transgression in the West if it violates someone's personal rights only because we have a worldview (and corresponding social practices) that imagines the individual in detachment from the community (the psychological social imaginary that was addressed in Chap. 2).

Even if Turiel's theory is superior to Kohlberg's in significant ways (especially concerning the developmental findings), it still suffers from the main problem with procedural theories: that the rules procedurally generated are abstract, sterile, and disconnected from our practical lives with each other. And, further, it is an unacknowledged defense of a Western worldview at the cost of portraying other worldviews as morally inferior.

Normative Practice Theory

As I have hinted, the label "practice theory" covers a motley group of theories that advance the imperative to "begin with practices!" in the analyses of human life (Stern, 2000). This imperative urges us to begin with what we *do*, instead of what we *think* about what we do. We should begin with life as concretely lived rather than with theories and reflections about lived life. This is not to say that "what we think about what we do" is an illegitimate object of knowledge, but when it does become an object of knowledge, practice thinkers insist that we examine it as yet another practical way of being in the world. Knowing about the world, theoretically and reflectively, is as much a practice as carpentry or flute playing. Knowing is something we do, and is to be understood as a practice rather than as detached theoretical representation.

In relation to the human and social sciences, the core idea of practice theory is that human and social life is not made up of isolated and individual actions. Actions can only be the kinds of actions they are because they are parts of larger wholes, that is, practices. The same action, for example, mowing the lawn, can also be part of different practices, for example, the practice of gardening, the practice of taking exercise, or the practice of acknowledging one's debt to the community. In a sense, then, the action is a different one when framed within different practices (although the physical movements may remain the same).[2] So actions are meaningful because they are parts of practices. "Any action presupposes an extended practice of carrying out that action," says practice theorist Schatzki (2000:100). Social reality simply *is* practices, as Charles Taylor argued in an early essay (Taylor, 1971).

[2] In Harré's (2002) terminology, the *action* is the same (mowing the lawn), but it represents different *acts* in relation to different discursive practices.

Rouse (2002) has argued that there are two ways of thinking about practices. One way identifies practices with regularities or commonalities. We thus have a practice when a number of people behave similarly across time. Accordingly, football and agriculture are practices because people tend to engage in these kinds of activity in ways that look alike. Often it is presupposed that there is a common structure behind these practices that regulate them (and this is certainly the case with the rules of football). Researchers influenced by structuralism in anthropology and sociology have set themselves the task of discovering such underlying, regulating structures.

The other view of practice understands practices as involving what Rouse calls *normative accountability* (Rouse, 2002:161). In this sense of practice, which Rouse sets out to defend, a practice may, but need not, involve a regularity or commonality. As he argues, "the normative accountability that binds a practice together need not even presuppose an uncontested formulation of norms to which all phenomena belonging to the practice are accountable" (p. 161). Rouse thus argues that although practices are normative in the sense that those actions and performances that make it up are accountable in relation to the practice, this does not necessarily imply that there are rules, procedures or norms *behind* the practice that function as that something relative to which we are accountable. Our performances are accountable relative to the practice, not to anything *behind* the practice, such as rules or transcendent ideas. For it may well be that there *are* no such rules or ideas – at least not before someone decides to formulate them. Pinkard describes a common ground between Hegel and pragmatists like Dewey on this point. Both articulate what Pinkard calls a "dynamic conception of normative authority" that tries "to anchor normative practice in the activities of *life* itself" (Pinkard, 2007:144–145).

In *Sources of the Self*, Taylor uses the word "practice" to deliberately denote "something extremely vague and general," as he says: "any stable configuration of shared activity, whose shape is defined by a certain pattern of dos and don'ts, can be a practice" (Taylor, 1989:204). According to this definition, we find the beginning of a normative conception of practices, for it tells us that those kinds of shared activity that do *not* involve a pattern of "dos and don'ts" are not to be thought of as practices. It is hard to come up with such examples, for – as the previous chapters hopefully have made it clear – human activities are saturated with normativity. But, to take a rather lame example, there are no dos and don'ts involved in determining whether to use the right hand or the left hand when opening doors. We simply use one hand or the other as it befits us, and we are never praised or blamed for our conduct in this regard. So there is no practice here, although there might well be some regularity or common forms of behavior (the majority of people probably use the right hand most of the time). Regularity is not enough to make a practice.

Aristotle

In a broad sense, the history of practice theory begins with Aristotle. Aristotle argued that by definition all human activity aims at some good, and cannot be understood in isolation from this good. Specific kinds of activity aim at specific kinds of end or good. The practices of medicine aim at health, military practices aim at victory, and economic practices aim at wealth (Aristotle, 1976:63). But there are, Aristotle observes, some activities that we do not engage in in order to achieve something beyond or outside the activity itself. We can say that I take piano lessons in order to learn the skill of playing the piano, but we normally cannot say that I play the piano in order to achieve something else. Playing the piano becomes a goal in itself. Of course, if I am a professional piano player, I may sometimes play merely in order to make a living, but if this were *always* the case, then we wouldn't say (at least Aristotle wouldn't) that I genuinely excelled in the practice of piano playing. Likewise, when I was taught to play the piano as a child (let's say), I did not do it for its own sake, but because my parents told me to. Again, there is a sense in which I did not really participate in the practice of piano playing (but rather in the practice of pleasing my parents). This is so because practices in Aristotle's sense are those activities that we engage in for their own sake, and not for the sake of something else. Consequently, if I always played the piano in order to make money, making money would be the goal of my activity and playing the piano would be an arbitrary means of reaching this goal. I would thus primarily be engaged in the project of making money, and *not* in the practice of musical performance. And, to complete this argument, making money can never be a practice in Aristotle's eyes, for money is a paradigmatic example of something instrumental, it is not its own end. For something to count as a practice (*praxis*) for Aristotle, "it must be undertaken as a *constituent* means to *eudaimonia* (that is, the agent's reason must be expressible on these lines: 'Doing this is what, here and now, doing well is')" (McDowell, 1998:7). But what, then, is *eudaimonia*?

Eudaimonia is a form of life that consists of actions and practices that we would choose for their own sake, and Aristotle defines the *eudaemon* – the person who lives well – as "one who is active in accordance with complete virtue, and who is adequately furnished with external goods, and that not for some unspecified period but throughout a complete life" (Aristotle, 1976:84). The emphasis on a complete life is very important in Aristotle's ethics, and it contrasts with modern utilitarian and Kantian conceptions of moral evaluation, which are primarily concerned with single actions: does this very action live up to pre-specified normative standards? In Aristotle's virtue ethics, it is primarily a whole life that is subject to moral evaluation, and only secondarily single actions within it.

According to Aristotle, the *eudaemon* is active "in accordance with complete virtue," and by "virtue" he means a disposition to act and feel in the right manner as demanded by the circumstances. Acting and feeling in the right manner has no exterior purpose, but is its own end. Due to the kind of creature that humans are, we have a certain function, and the virtues are those capacities and dispositions that enable us to realize our function and live a form of life that is good and proper for humans – and thus be *eudaemon*. Aristotle is a naturalist and thinks of humans as he would think of swans or bees. In virtue of the kind of creature a bee is, there are certain things that it ought to do. If a predator attacks the beehive, the bee ought to counter-attack, sting, and consequently die. If the bee is a coward bee, and prefers to suck nectar instead of defending its hive, we are allowed to say that it is defective, and does something normatively wrong (Philippa Foot has recently developed this view, see Foot, 2001). Defending the hive would be a practice for the bee, something that is part of what we consider a proper form of bee life, and bee strength and "courage" would be virtues that enabled the bee to excel in the practice of defending the hive. But in order to know this, we need to know about bee teleology, that is, about what a perfected bee life looks like. Aristotle would argue that the same applies in the human realm: In order to know what practices are, that is, know which actions are done for their own sake, and know what those virtues are that allow us to excel in human practices, we must know about *eudaimonia* or the human *telos*, the perfected state of human life, and also about the human *energeia*, the distinctively human mode of functioning (Lovibond, 1995).

This point is rather alien to modern moral theory. It involves the view that there are certain natural normative and substantive standards as to what constitutes a life well lived. Our preferences, desires or feelings are not the final arbiters in this matter, although an important part of moral life concerns the formation and cultivation of human sentiments through the course of upbringing. We must learn to love, hate, admire and fear the right things. This we do by acquiring a "second-nature"; habitual modes of discernment and responding to the moral order and "the space of reasons" that are attained through education, socialization, and relationship (Reader, 2000:343). As opposed to proceduralism, it involves the view that certain natural and worldly standards determine moral life and not anything that transcends empirical realities such as abstract procedures. It is the *eudaemon* that functions as the standard of 'good life,' and the *phronimos* (the practically wise person) who functions as the standard of correct judgment. Aristotle would not be against moral rules, but he would be for them only to the extent that they assist humans in approaching their proper form of life. He would see procedures and rules as tools, rather than authorities, concerning morality. What *are* authoritative are those substantive values, in the form of valuable activities comprising a flourishing life, which humans engage in without exterior purpose. But how does Aristotle then define human perfection and the human function?

The *Nichomachean Ethics* is all about this complex question, but the brief answer is that he points in two different directions. Both directions, however, are related to reason because reason is the distinct mark of the human, according to Aristotle. We are rational animals with the capacity for conceptual thought and thus with the capacity to respond to meanings and reasons. We are creatures that are able to give reasons for what we do, and expect such reasons in return from others. So the human function is the exercise of our rational and discursive powers, which, of course, have to be cultivated in and by the communities of which we are a part. But Aristotle here points towards two kinds of rationality, which is something that has confused many exegetes. On the one hand, his *Ethics* culminates in the final book ten with the assertion that contemplative activity is the highest form of activity, something we do for its own sake, and thus is constitutive of *eudaimonia*. So the perfected kind of life is one that involves the ability to theoretically contemplate our place in the cosmos, something that contributes to nothing else but the act of contemplation itself. In the preceding books, however, much more emphasis is put on the practical and social aspects of life, on our capacity for practical rationality in a community of practical reasoners, and some have even wondered whether the final book ten could be an addition made by one of Aristotle's students. For in the other parts of the *Ethics*, "political" activities are at center stage.

"Political" comes from the Greek word for city, *polis*, and Aristotle famously claimed that the human being is a political animal, a *zoon politikon*, an animal that can live a proper form of life only in the socially organized context of a community. A human being detached from her community is unable to exercise practical rationality: "the *polis* is required for *arete* and for *phronesis* [...]. Separated from the *polis*, what could have been a human being becomes instead a wild animal" (MacIntyre, 1988:98). Indeed, the *polis* is "the locus of rationality" (p. 141), and the conceptions of the human being as a rational and political animal are therefore two sides of the same coin. All practical rationality is political rationality in Aristotle's scheme. Political science is thus more fundamental than ethics, for the latter is merely concerned with individual character, and not with overarching social goods. For Aristotle, "political science" (the science of the city community) is "the most authoritative and directive science" (Aristotle, 1976:64), for it studies the supreme good for humans. And it not merely studies it passively and at a distance, so to speak, for political (and also ethical) knowledge in Aristotle's sense involves the ability to *act* well and responsibly among other human beings in one's community.

Aristotle says that "if anyone wants to make a serious study of ethics, or of political science generally, he must have been well trained in his habits" (Aristotle, 1976:67). A sort of practical, habitual training in feeling right and acting well should precede a theoretical instruction in the proper form of life (ethics) and the proper form of community (politics). One must already *be* good, to a certain extent, in order to be able to grasp theoretically what the good *is*. In order to reflect upon, and possibly renew one's community and its moral traditions, one must have acquired certain bodily, communicative, and social habits beforehand (Burkitt, 2002; van Alstyne, 1998). In a way, the practical sciences (ethics and politics) are merely theoretical reminders that serve to remind the good person of what she already knows and has incorporated in her life. The point is that practices are primary in moral life, and that moral theory makes sense only given the background of already inculcated moral habits and settled dispositions.

Moral Goods and Practices: MacIntyre's Contribution

MacIntyre's *After Virtue* (1985a) has been one of the most influential works in moral philosophy – and related disciplines – within the last 25 years or so. It represents a thorough attempt to rehabilitate Aristotelian virtue ethics, but in a way that does justice to the fact that the virtues today are understood as much more heterogeneous and historically dependent than in Aristotle's times. Although it is a historicized version of virtue theory, MacIntyre denies that it is a relativistic version.

In a recent encyclopedic article on virtue ethics, MacIntyre argues that in Greek philosophy, all ethics was virtue ethics (MacIntyre, 2001). This is so because a virtue was an excellence (*arete*), and ethics was the practical study of excellences of character. In *After Virtue*, MacIntyre understood the history of Western moral philosophy to be one in which philosophers (and laymen) gradually "forgot" that all ethics was virtue ethics, and this eventually resulted in what he calls the failure of the Enlightenment project of justifying morality. MacIntyre's historical narrative roughly runs as follows: Originally, in the Greek world, ethics was part of a teleological scheme in which there was "a fundamental contrast between man-as-he-happens-to-be and man-as-he-could-be-if-he-realized-his-essential-nature" (MacIntyre, 1985a:52). Ethics was, as Macintyre recounts, the practical science of how to make the transition from the former to the latter state. Ethics thus presupposed an account of the human *telos*, the perfected end state, as we also saw in Aristotle's ethics. Practical reason is the capacity that enables us to make this transition in the course of a human life. Practical reason is thus, in the way I have used the term above, "substantive," because it makes sense only given certain substantive conceptions about the contents of a life well lived (man-as-he-could-be-if-he-realized-his-essential-nature).

With this book's Chap. 5 in mind, I hope the reader will realize that modern objections to this idea that it commits "the naturalistic fallacy" are off the mark. For "functional concepts," such as a watch or a farmer, we can indeed correctly infer evaluative judgments from factual descriptions. From "this watch is inaccurate and too heavy to carry about" we may validly conclude that it is a bad watch. This is so because a watch *is* something that *ought* to tell time accurately and be handy enough to be carried about. We cannot understand what a watch is, if we do not know its proper function, because what a watch is, is defined in relation to this function. Therefore, to call a watch good is always also to make a factual statement. It is to say that it is what it ought to be; or does what it ought to do. In the Aristotelian tradition, "human being" is just such a functional concept, and we cannot understand what a human being *is*, according to this tradition, without an understanding of how the human being *ought* to live.

Two things happen in Western history that problematize or even destroy this Aristotelian view. First, as I have touched upon in the first chapters of this book, the individual is born as a psychological entity in Western history. This contrast with earlier times and traditions where, as Macintyre says:

> To be a man is to fill a set of roles each of which has its own point and purpose: member of a family, citizen, soldier, philosopher, servant of God. It is only when man is thought of as an individual prior to and apart from all roles that "man" ceases to be a functional concept (MacIntyre, 1985a:59).

When the individual becomes the primary unit of social life in the modern psychological social imaginary, it is no longer possible to uphold a functional or teleological conception of the human being. For each individual is imagined to exist prior to and in abstraction from the context of community and tradition. Therefore, it has seemed to philosophers, psychologists, and social scientists that to call someone good is not at all to make a factual judgment. This would presuppose that humans were defined by their functional roles, rather than by their more abstract individualities. We have seen that psychology entered the scene when the individual (conceived as an entity that transcends its social roles) was born. So it is no accident that psychology has by and large accepted a strict separation of factual and evaluative judgments when describing and explaining phenomena in human lives. This (imagined) separation was the soil out of which psychology grew, so to speak.

The other thing that destroys the functional conception of the human being is the rise of mechanistic science. Aristotle's teleological worldview, where things were understood in terms of their functions and goals, was superseded by the modern, disenchanted view with great scientists like Galileo and Newton. Now, things are no longer understood in terms of intrinsic forces that develop in accordance with predetermined ends, but in terms of how other things affect them. This view was also carried over into the social and human sciences, psychology among them, and the possibility that human and moral inquiry could and perhaps should work in different ways was ignored. As Charles Taylor says:

> The notion that human beings have something like a telos *qua* human being can be separated from the thesis that everything in nature belongs to some class or other, whose behavior is explained by some Form or Idea. Because we no longer explain the movement of the stars and stones teleologically does not mean that we cannot explain humans in these terms (Taylor, 1994:17).

If we go back to MacIntyre's narrative of Western moral philosophy, we can note that he reads its history in the light of the repudiation of the older teleological view of humans. What happens is that a crucial element of moral life disappears: "man-as-he-could-be-if-he-realized-his-essential-nature." All that philosophers were left with then were "man-as-he-happens-to-be" and the rules and principles of practical reason, but *without* the substantive teleological context in which these rules and principles originally belonged and were rendered meaningful. How can morality then be justified? This has been the leading question in modern moral philosophy. The logical and historical consequence has been emotivism, the view that there simply is no way to justify morality, for, in a way, there is no morality. There are only emotional exclamations of individual preference; thus "this is good" means (in C. L. Stevenson's sophisticated emotivism) "I approve of this – do so as well," or (in A. J. Ayer's cruder version), it simply means "hurrah for this!" In MacIntyre's history of moral decline, this is what we are left with today, and, as he says, emotivism has been embodied in our institutions, notably in the form of bureaucracies, and in our *wertfrei* social sciences, *and*, of particular interest to psychology, in therapists who incarnate the emotivist logic according to which only means can be rationally discussed, and where ends can only be determined by individual preferences (see also Bellah, Madsen, Sullivan, & Tipton, 1985).

Fortunately MacIntyre does not leave us with this gloomy scenario. He is interested in retrieving a conception of the human *telos* that on the one hand reinstalls the substantive context that renders morality meaningful and on the other hand does not suffer from what he sees as the ills of Aristotle's "metaphysical biology."[3]

---

[3]In a later book, *Dependent rational animals* (MacIntyre, 1999a), MacIntyre is much more open to a non-historicized version of virtue theory that is grounded in our biological, animalistic nature. I believe we should read this as an *addition* to the account of virtue developed in *After Virtue* rather than as a contradiction.

MacIntyre develops his account of the human *telos* through an understanding of the virtues, because the virtues are those well-formed capacities and dispositions that enable us to live according to our *telos*. He believes that there is "a unitary core concept" of the virtues (MacIntyre, 1985a:186) of which he gives an account via three concepts: practice, the narrative order of a single human life, and a moral tradition. I shall here briefly explain this stage-wise development (MacIntyre thinks that each stage presupposes the earlier, but not vice versa).

Practices

First, as MacIntyre makes clear, virtues are situated within practices. There can be no such thing as human excellences if there is nothing to be excellent *at*. And what we can be excellent at are practical activities. Football, chess, architecture, farming, science, history, painting and music are mentioned as examples of practices (MacIntyre, 1985a:187), and the strict (yet complex) definition given of a practice is:

> Any coherent and complex form of socially established cooperative human activity through which goods internal to that form of activity are realized in the course of trying to achieve those standards of excellence which are appropriate to, and partially definitive of, that form of activity, with the result that human powers to achieve excellence, and human conceptions of the ends and goods involved, are systematically extended (MacIntyre, 1985a:187).

As with Aristotle's conception of practice as something done for its own sake, MacIntyre here defines a practice as a form of activity with internal goods. If there are no goods internal to an activity, i.e., goods that I cannot achieve without engaging in this form of activity, then the activity does not qualify as a practice. I have already tried to explain what it means to say that goods are internal to activities, when I discussed the piano-playing example. It simply means that the excellent participant in the practice (although not necessarily the apprentice) will perform the required actions for their own sake, and not for the sake of anything external to the practice, such as money, power, or fame. Such external goods (money, power, fame) are always some individual's possession, MacIntyre notes (1985a:190), whereas the achievement of goods internal to practices contributes to the whole community that participates in the practice. The artist who excels in the practice of portrait painting does not take anything away from other portrait painters, but indeed advances the whole practice itself. How to excel in a practice is not up to individuals, but to the standards of the practice. This has important implications for how to become an excellent participant in a practice: "we cannot be initiated into a practice without accepting the authority of the best standards realized so far" (p. 190). The only way to learn is to accept one's own initial incapacity, for the "goods can only be achieved by subordinating ourselves within the practice in our relationship to other practitioners" (p. 191).

From his definition of practices as those activities that have internal goods, MacIntyre constructs an updated account of the virtues:

> A virtue is an acquired human quality, the possession and exercise of which tends to enable us to achieve those goods which are internal to practices and the lack of which effectively prevents us from achieving any such goods (MacIntyre, 1985a:191).

Virtues are the human qualities that enable us to achieve practice internal goods. Thus, objectivity, openness, patience, risk-taking, perseverance and honesty may be virtues that enable the practitioners of science to achieve the goods internal to scientific practice. Other forms of practice may demand other virtues. MacIntyre argues that his view of practices and virtues avoids Aristotle's problems: "although this account of the virtues is teleological, it does not require any allegiance to Aristotle's metaphysical biology" (MacIntyre, 1985a:196). It is primarily a cultural and historical view of virtues, and only secondarily a biological one. Further, MacIntyre adds, it portrays practices and goods as variegated, incompatible, and multiple, which means that "conflict will not spring solely from flaws in individual character" (p. 197). There is bound to be conflict and choice in individual lives, because we live in and with a large number of practices that we have to balance against each other. We are always participants in more than one practice, and conflicts arise because of a clash between practices, and not (necessarily) because of a collapse of individual human character.

The Narrative Order of a Single Human Life

Although practices are primary as the arena in which virtues are situated, an account of morality and moral life must surpass our participation in single practices. We need to know what it means to participate *across* practices in the course of a lifetime. If *everything* moral were situated in practices, there would be too many conflicts and too much arbitrariness in our lives (MacIntyre, 1985a:201). Thus MacIntyre needs an account of a whole human life as a unity, for "without an overriding conception of the *telos* of a whole human life, conceived as a unity, our conception of certain individual virtues has to remain partial and incomplete" (p. 202). We need to know what the right kind of life looks like that allows us to participate adequately (at the right times, in the right ways, in the proper amounts) in different practices. Practical wisdom (*phronesis*) is the ability to judge concerning what to do when different practices set up different and perhaps incompatible ends. In such situations, we cannot appeal singularly to any of the conflicting practices, but must use our practical and particularistic powers of judgment.

What do we appeal to? If not to practices, then to what else do we appeal? It is here that MacIntyre develops his account of the narrative unity of a human life as the standard in the light of which we can navigate when practices clash. This fits nicely with Aristotle's emphasis on the whole life as the subject of moral evaluation. MacIntyre notes that "Narrative history of a certain kind turns out to be the basic and essential genre for the characterization of human actions" (MacIntyre, 1985a:208). Actions are not only situated in the context of practices, but also in the context of a human life; that is, actions are at once constitutive of, and constituted by, the narrative lived by the person.[4] "Stories are lived before they are told," says MacIntyre (p. 212), but the way they are lived is not just up to individuals. We are not sole authors of the narratives of our lives, for "We enter upon a stage which we did not design and we find ourselves part of an action that was not of our making" (p. 213). And further:

> The key question for men is not about their own authorship; I can only answer the question "What am I to do?" if I can answer the prior question "Of what story or stories do I find myself a part?" We enter human society, that is, with one or more imputed characters – roles into which we have been drafted – and we have to learn what they are in order to be able to understand how others respond to us and how our responses to them are apt to be construed (MacIntyre, 1985a:216).

MacIntyre's account does not, therefore, merely point towards practices, but towards what has come to be known as *communities of practice* in anthropological writings (e.g., Lave & Wenger, 1991). It means that in our life narrative we are "condemned to meaning" to use an expression from Merleau-Ponty (1945:xxii). We are "thrown" (cf. Heidegger, 1927) into the community's structures of narrative significance that we have to acknowledge and make use of if we are to make sense. We are "thrown" into the space of reasons. MacIntyre's view implies that "the characters in a history are not a collection of persons, but the concept of a person is that of a character abstracted from a history" (MacIntyre, 1985a:217). The modern view of individuals, persons, or selves as something primary in history and social life is pure fiction. In reality, we live our lives according to scripts that are drafted beforehand. The practices in which I am engaged – as a father, husband, academic, etc. – are not of my own making, and the ways I orchestrate my participation in such practices owe much to narrative forms that precede my own life. This, however, does not mean that humans are not responsible or accountable:

> To be the subject of a narrative that runs from one's birth to one's death is [...] to be accountable for the actions and experiences which compose a narratable life. It is, that is, to be open to being asked to give a certain kind of account of what one did or what happened to one or what one witnessed at any earlier point in one's life than the time at which the question is posed (MacIntyre, 1985a:217f).

To be someone – to be a "subject" and to have an identity – means to be accountable, and accountability is indeed a fundamental moral virtue across practices. Accountability involves the ability to place one's actions in "the space of reasons," thereby rendering them intelligible, and human subjectivity thus means responsiveness to (moral) reasons. This is in line with a central thesis of this book: That mental life is lived in the space of reasons.

MacIntyre (1999b) has argued that the two virtues of integrity and constancy are necessary in order for us to exercise other moral virtues. One must be the same person across contexts and practices (integrity) and pursue the same goods through

---

[4] The notion of the hermeneutical circle is relevant here. The whole – human life – is made up of actions, which again derive their meanings from the larger life narrative. MacIntyre's narrative view of self and identity closely resembles that of Ricoeur (1992).

extended periods of time (constancy) if one is to be an accountable and trustworthy moral agent. It makes no sense to be asked to give "accounts" of one's actions without presupposing integrity and constancy. Taylor (1989) correspondingly refers to the unity of a whole human life as the *teleion agathon*, the single complete notion of what is good that determines the place and proportions of other, more partial, practice-internal goods. The unity of a human life is what enables us to rationally combine the different practice internal goods in a single life. It is an "architectonic" good (Taylor, 1994) because it concerns the *shape* of our lives, and awareness of such a good helps us confront and choose between incompatible moral reasons for action (Brinkmann, 2004b:67).

A Moral Tradition

The fact that lives are lived in a narrative mode on the background of pre-existing structures of significance means that emphasis has to be put on the notion of tradition. Something is a practice, and something is a narrative, only as part of a moral tradition. And "tradition" therefore is the third core concept in MacIntyre's reconstructive account of virtue. Practices and narratives are "embedded in and made intelligible in terms of the larger and longer history of the tradition through which the practice in its present form was conveyed to us" (MacIntyre, 1985a:222). Thus, it becomes a virtue in itself to have an adequate sense of the tradition(s) to which one belongs. MacIntyre has been criticized on this point and has been accused of moral conservatism. Before dismissing his position as outdated conservatism, however, we should carefully examine the way he approaches the notion of a tradition. "A living tradition," he says, "is a historically extended, socially embodied argument and an argument precisely in part about the goods which constitute that tradition" (p. 122). A tradition is thus not something static and oppressive, but rather an ongoing discussion that extends through generations about what is worthwhile and what is not. This seems to me to be not conservatism, and if it is, it is a kind of conservatism that is hard to escape if one wants to remain intelligible. For every discussion, and every moral inquiry, must inscribe itself in that history of which it necessarily is a part (even anarchism or moral nihilism are intellectual traditions). MacIntyre's view of traditions resembles Gadamer's (1960), and, like Gadamer, he does not see traditions as oppressive forces that limit individual freedom and autonomy, but rather as enabling factors that make possible any human practical and moral inquiry.[5]

[5] Kant's (1781) image in his first *Critique* of the light dove, which cleaves the air in her free flight and feels its resistance and imagines that its flight would be still easier in empty space, comes to mind. Its flight wouldn't be easier in empty space, because there it wouldn't be possible, and likewise human life wouldn't be freer without traditions and practices, because there it wouldn't be possible!

Reasons and Articulation of Practices

We can now see emerging a forceful alternative to the dominant accounts of practical rationality given by the instrumentalist and proceduralist traditions. For instrumentalists, the end is given ("preference satisfaction" or some variety of this), and our practical rationality is merely concerned with devising effective means of reaching this end. A reason for action is thus a good reason if it lives up to, and directs us to, the predetermined end. In the proceduralist tradition, on the other hand, a good reason for action (a moral reason) is when my action is based on the principle of practical reason: for example, the Kantian categorical imperative.

In the practice-based approach, however, reasons for action become relative to practices, that is, relative to activities that have internal goods. Thus, the practice-based view represents a substantive conception of practical rationality. In this approach, a moral reason can be said to be "any consideration which does not conflict with the overarching good of human life, against which other ends are measured" (Reader, 2000:355f). Reasons are thus related to the substantive human good in the right way. Virtues, on this account, are not themselves reasons for action,[6] but they are capacities to *recognize* reasons (p. 362) as they arise in ongoing practices. Thus, reasons explain, justify and rationalize (in the sense of demonstrate why it was rational to act in this way) relative to practice internal goods (and also to those goods that transcend practices, as we have seen). What it means to be a human agent, on this account, is to be able to respond to, articulate, and ask for reasons.

Why begin with practices, MacIntyre asks in the postscript to the second edition of *After Virtue* (1985a:273)? As he says, other views of morality and practical rationality have begun "from a consideration of passions or desires [the tradition of Hume and instrumentalism, SB] or from the elucidation of some conception of duty or goodness [the tradition of Kant, Kohlberg and other proceduralists, SB]" (p. 273). The main problem with these traditions lies in their inadequate view of means and ends "according to which all human activities are either conducted as means to already given or decided ends or are as simply worthwhile in themselves or perhaps both" (p. 273). What this view leaves out "are those ongoing modes of human activity within which ends have to be discovered and rediscovered, and means devised to pursue them; and it thereby obscures the importance of the ways in which those modes of activity generate new ends and new conceptions of ends" (p. 273). These modes of activity in which new ends are generated are what MacIntyre has referred to as practices. Concerning the virtues, both Aristotle and MacIntyre treat them as means to an end in a sense, viz. as means to achieve the goods internal to practices,

[6] If I justify an action by saying that my reason for action was to let my action express a virtue, then it nearly disqualifies the action as a moral action, for a moral action ought to be concerned with the *object* of the action (e.g., to help someone), rather than with the quality of the agent (Reader, 2000).

but what distinguishes this view from a modernist separation of means and ends is the fact that "the relationship of means to end is internal and not external. I call a means internal to a given end when the end cannot be adequately characterized independently of a characterization of the means" (p. 184). There is an internal relationship between virtues and practices. It becomes clear on this view of practices that new ends can arise within practices, and consequently new virtues. In this sense, there are no completely fixed ends. Thus, given this updated practice oriented approach to virtue ethics, it becomes possible to answer some of the cogent criticism that has been directed at it in Western philosophy.

One such critical point comes from John Dewey. Dewey was always arguing against any conception of life that operates with fixed ends. Dewey wanted to move our preoccupation in moral life "from following rules or pursuing fixed ends over to the detection of the ills that need remedy in a special case and the formation of plans and methods for dealing with them" (Dewey, 1920:165). If we take this pragmatic move, "The end is no longer a terminus or limit to be reached. It is the active process of transforming the existing situation. Not perfection as a final goal, but the ever-enduring process of perfecting, maturing, refining is the aim of living" (p. 177). Given the above interpretation of practices and virtues, it in fact does become possible to operate with a notion of perfection (or perhaps *perfecting*, as Dewey would have it), for the virtues are those capacities that enable us to detect reasons for action arising from goods internal to practices, and this is a never-ending process. So the account developed above, drawing on Aristotle and MacIntyre, can agree with Dewey in that means and ends are not contrary to each other, but should be seen as a continuum, and yet insist that there has to be at least one "fixed" end: To be sensitive and responsive to the moral reasons that there are in our practical lives. What these reasons are, more specifically and concretely, can only be understood by drawing in the actual practical situation at hand. This way of putting it – responsiveness and sensitivity to reasons for action – owes much to John McDowell's work, where virtues are conceptualized as "specialized sensitivities to requirements" (McDowell, 1998:52). What I have wanted to do in this chapter is supplement this understanding with the insight of practice theory that these "requirements" are either embedded in practices, or appear when practices are in conflict or break down.

Rationality as Articulation of Practices

A distinct approach to human rationality thus emerges in the practice-based interpretive-pragmatic view. Charles Taylor has done much to develop this. He calls it *rationality as articulation*. Both instrumentalism and proceduralism see rationality as the employment of some method (e.g., the utilitarian calculus or the categorical imperative). Here the methods determine what we can understand, and how we can understand it. In contrast, the practice-based view understands rationality as articulation of what is inherent in social practices. Taylor draws on arguments from Heidegger, Merleau-Ponty, and Wittgenstein (and could have included Dewey) to demonstrate that the procedural and instrumentalist attempt to find universal methods with which to "turn the background against which we think into an object for us" is misguided (Taylor, 1995c:12). Taylor argues that it makes sense to employ methods only when given a background that is comprised of practices that can never be made totally transparent and explicit. This they cannot, because, at bottom, our understanding is embodied and rests on habits (Dewey, 1922) or what Bourdieu (1977) called *habitus*. There are certain contextual background features that determine when and how to use which methods and these features cannot be made fully explicit. Taylor distinguishes between three levels of understanding: Embodied understanding ("background understanding"), symbolic understanding (expressed in ritual, symbols, and works of art) and explicit understanding (Taylor, 1999:167). And he argues that the embodied level is primary. But if we cannot turn the background completely into an object for us with the help of methods, then how should we conceive of rationality?

> The task of reason has to be conceived quite differently: as that of articulating the background, "disclosing" what it involves. This may open the way to detaching ourselves from or altering part of what has constituted it – may, indeed, make such alteration irresistible; but only through our unquestioning reliance on the rest (Taylor, 1995c:12).

Our understanding of the world is grounded in our practical dealings with it; dealings that are structured (or, if this is the wrong word, *lived*) in social practices. Understanding is *not* primarily based on representation, for there are always contextual (practice-based) conditions for anything to count as something (Dreyfus, 1991; Heidegger, 1927; Taylor, 1995b).

The Nature of Psychology

"What does this have to do with psychology?" one might ask. How are these distinctions and complex accounts of virtues and practices relevant to psychology and other human and social sciences? My argument is that they are relevant because psychology is concerned with human action and if so, it follows from the present chapter that it has to begin with those practices of which actions are parts. And if the normative view of practices has some plausibility, it follows that we cannot conceive of practices without an understanding of the moral values and goods that are embedded in them, and in fact serve to define them as practices. Thus, if psychology wants to understand human action, then it cannot do so without an understanding of moral values and goods. Contrary to its dominant traditions, psychology has to incorporate a normative and moral framework in its understanding of human life if it wants to remain faithful to the phenomena at hand.

To summarize: Actions cannot be understood independently of practices, and practices cannot be understood independently of excellent participation in practices, because this is what defines practices as such (this is what Aristotle hints at when he said that the excellent practitioner is the normative standard of right living). Psychology should thus take an interest in humans as participants in practices, and in those capacities that humans must have in order to excel in practices (viz. virtues). In my view, psychology could here learn much from virtue theory. As I read MacIntyre (which, I think, is also the way he would *like* to be read), his work is not *merely* philosophy, but obviously also philosophical history *and* social history, and *also* a contribution to sociology and psychology as such. I believe we can use his approach not just as inspiration for psychology, but in fact as a kind of normative psychology in itself. That is, a kind of practice-based psychology that begins from practices, moral goods, narratives, and virtues in understanding human action, rather than from value neutral mental representations, beliefs and desires, or information processing. I would like the present chapter to be read not just as philosophical clarification of certain issues in psychology, but as laying the foundations for psychological inquiry *qua* psychological, that is, for a psychology that takes seriously the idea that mental life is lived in a moral space of reasons.

Conclusions

My intention in the present chapter has been to use the ideas of Aristotle and MacIntyre to develop a vocabulary for the interpretive-pragmatic view in which to conceptualize moral phenomena in psychology. This vocabulary has primarily situated morality and values in practices. Practical reasoning is here *not* first and foremost the capacity to employ prefigured methods of reasoning but *is* the active search for moral reasons for actions as they arise and are displayed in and between ongoing social practices. I have defended a neo-Aristotelian view according to which virtues are situated within practices. The correct reply to the question "Is he good?" would here not be concerned with the issue whether the person in question follows certain methods, but pose the counter-question "at what?" For being 'good' means doing something well, excelling in some practice defined by normative standards, fully understandable only with a vocabulary of what I have earlier referred to as thick ethical concepts. I believe this way of proceeding solves some of the problems in instrumentalist and proceduralist models of moral reasoning. Against the former, it conceptualizes sentiments, emotions, and desires not as moral authorities, but as something that should be continually evaluated normatively in the light of practice-based values and goods. It thus preserves our everyday intuitions about the normativity of feelings. Against the latter, it points to the notion that principles and procedures have no moral authority in themselves, but have authority only given a background of practices and their substantive values.

We are left with one significant problem: As MacIntyre tirelessly remarks, we live in a culture in which practices (in the normative sense) are endangered. Dreyfus even refers to practices in the normative sense as "marginal practices," and argues in an interview with Bent Flyvbjerg that they are threatened by an instrumental culture (Flyvbjerg, 1991). Dreyfus refers to friendship as a marginal practice that so far has resisted attempts at instrumentalization and rationalization. But even friendship might be threatened, for example, by a reduction to "networking."

The point being that the context in which practices were ordered (the *polis*) no longer exists, which means that we must look elsewhere for moral guidance. This points to a possible problem in the practice-based account behind the interpretive-pragmatic view: The Aristotelian perspective may find it hard to account for morality in a cultural situation without communities and agreed-upon conceptions of human teleology. It seems to me, however, that there is still a minimal and necessary *telos* on which we all (implicitly) agree, even in a pluralistic age, viz. receptiveness to the reasons that there are. Any psychological conception that excludes the human capacity for responding to such reasons is defective (because it is this very responsiveness that is constitutive of the mind). We may be more Aristotelian than we think, as Taylor (1994) has argued. Although we no longer have a single "superstructure" like the *polis* to order practices, we still have a number of practices that seem worthwhile to engage in.

Chapter 8
Conclusions

In this concluding chapter, I shall do three things: (1) State the main conclusions that I have drawn in the text and summarize the main tenets of the interpretive-pragmatic outlook on psychology and morality. (2) Raise some relevant objections and challenges to the interpretive-pragmatic framework, which will hopefully elucidate some problematic points and suggest things to work with, in the future. (3) Briefly analyze some central psychological concepts in light of the interpretive-pragmatic framework to illustrate how these concepts can be reinterpreted as moral concepts. This analysis can be considered as a kind of research menu for further theoretical and empirical investigations of psychology as a moral science.

Two Main Conclusions

This book has examined some ways in which psychology – the discipline, its subject matter, and its problems – are related to moral issues. I have argued that the relationship between psychology and morality are so pervasive that it warrants dissolving a rigid distinction between them: Psychology is a moral science and psychological phenomena are moral phenomena. It should be borne in mind that I have said little about the reverse perspective, i.e., whether moral phenomena are necessarily also to be thought of as psychological phenomena. I would hesitate to draw this reverse conclusion since one of my main points has been that morality is "older" than psychology – from a cultural and historical perspective (cf. Chap. 2) – and it seems possible indeed to engage in moral inquiry and action without the institution and vocabulary of modern psychology. I have even argued that modern psychology, in so far as it has come to exclude normative issues from its field of operation, has made it quite difficult for us 'moderners' to comprehend the moral features of our lives. So any psychological issue *qua* psychological is a moral issue, but any moral issue is not necessarily a psychological issue (think of problems in bioethics, religious ethics, and politics for example). I have argued that we should think about psychology as *one* (often quite problematic) way of ordering the normative moral domain – summarized as the psychological social imaginary – but there are no doubt other ways. Ethics, politics, religion, esthetics,

S. Brinkmann, *Psychology as a Moral Science: Perspectives on Normativity*,
DOI 10.1007/978-1-4419-7067-1_8, © Springer Science+Business Media, LLC 2011

logic, and other normative fields of inquiry cannot be reduced to psychology, although psychology is, in its own ways, a normative activity.

The central point has thus been that it is fruitful to think of psychology as a moral science, and two main conclusions have led me to this:

1. Psychology is a value-laden social intervention that contributes to forming our self-understanding. Psychology makes up people according to certain implicit value systems. Psychology is a historical process with *moral effects*. This was treated mainly in part I.
2. Psychological phenomena are moral phenomena. Psychological phenomena can only be understood from an evaluative standpoint (e.g., we can comprehend human action only if we master thick ethical concepts). Psychology has *moral contents*. This was treated mainly in part II.

Psychology's Moral Effects

The first part of the book tried to describe, analyze, and propose reasonable reactions to the way psychology has worked, and continues to work, as a social and historical intervention, i.e., as a science that is involved in the constitution of its subject matter. In Chap. 2 I described a modern psychological way of thinking about human life, subjectivity, and morality with roots in modernism (Hume) and romanticism (Rousseau) that I referred to as the psychological social imaginary. Then, I gave a historical exposition of the changing notions of subjectivity and of the rise of different psychologies that have influenced our social imaginary and the moral order. In Chap. 4 I analyzed the processes that enable psychology to make up people, and I discussed two competing ways of comprehending these processes, advocating primarily an interpretive account over a Foucauldian one.

Rather than rejecting psychology as a science and practice after having seen that it is involved in "making up people," we should make the most of this and thereby hopefully gain in real-world utility and insight (Bredo, 1998:463). My overall argument in the first part was that a science that recognizes (rather than ignores) its own function as a social intervention must, if it wants to be both morally useful and to adequately understand itself, be conceived as a moral science. I have presented the science and practice of psychology as part of a larger historical development, which culminated in the psychological social imaginary; a cultural situation where people primarily think of social life in psychologically informed ways, viz. as dependent on individual private selves, an inner world, and social atomism. Psychology is at once the offspring of the psychological social imaginary and its most significant promulgator. In Chaps. 2–4, psychology was not just presented as a *social* intervention, but also as a *historical* intervention, which began at a certain place and time (and which may possibly be replaced by other modes of understanding). As Danziger has put it:

> Before there could be anything for the discipline of psychology to study, people had to develop a psychological way of understanding themselves, their conduct, and their experiences. They had to develop specific psychological concepts and categories for making themselves intelligible to themselves (Danziger, 1997b:139).

I have tried to show that psychology has been a particularly productive science, one that has produced a number of distinctions, categories, concepts, techniques, and practices, which have thereby fabricated what Hacking calls '*human kinds*'. My main interest in outlining the psychological social imaginary and its human kinds has been normative. I have tried to describe and analyze how the psychological social imaginary often has gone hand in hand with theories that render morality subjective, i.e., attempts to *psychologize* morality (and normativity more broadly), as I have put it. Psychology emerged in a culture that could not easily think of moral values as real aspects of the world, for the disenchanting natural science of the day had "demonstrated" that the world contained only facts. Values were then confined to the newly conceived inner subjective realm of the mind that psychologists purported to study. The result has been that psychology's social interventions have often been carried out in the name of science, value-neutrality, instrumentalism etc. According to instrumentalism, subjectivism, and emotivism, the end of human action – preference satisfaction – is given, and only the means of how to reach the end is up for rational discussion. And here, psychology has offered us a range of practices and techniques through which autonomous men and women can find the best ways to realize their preferences (and realize their *selves*), but only rarely have psychologists discussed how to evaluate if such preferences are in fact *worth* realizing. They have confined themselves to causal explanation and instrumental techniques rather than normative discussions of the justifiability of the reasons that people have for doing things.

In this, they not only risk unwitting participation in the "emotivist culture" of the West, but they also ignore those features of human life that ought to be foundational in any study of the mind as such. This takes me to the second major issue that has been addressed in this book: That psychological phenomena are moral phenomena, or, as John McDowell has said, "that mental life is lived in the space of reasons" (McDowell, 1998:296).

Psychology's Moral Contents

As argued throughout the text but mainly in the second part, psychological phenomena differ from other sorts of phenomena (chemical, physiological, etc.) in being subject to moral evaluation. There are normatively correct (and incorrect) patterns of thinking, feeling, and acting, and, as we saw in Chap. 7, these patterns can fruitfully be thought of as embedded in social practices[1]. They neither arise from subjective feelings and

[1] A relevant objection here is that it is premature to equate normativity with *moral* normativity. This is true in a sense, for it is not as such a *moral* defect, for example, to be unable to follow the normative rules of logical deduction. However, when we are concerned with human action, feeling, and thinking in real life contexts, we cannot make any hard-and-fast distinction between moral and other kinds of normativity. Here we are in the realm of *ethics* in the Greek sense, concerned with the proper way of life for humans. Here, it seems to me, normativity is also *moral* normativity. The modern sense of morality as a distinct sphere of life was wholly alien to the Greeks, and, as MacIntyre has argued, a sphere of morality in separation from human practices as such is chimerical.

sentiments (as Humeans and sociobiologists argue), nor from universal procedures of rational deliberation (as Kantians argue). Normativity demands the existence of practices, i.e., temporally extended ways of doing things, achieving goals and cooperating. In Chap. 7, I defended a normative notion of practices that goes back to Aristotle, according to which practices are those modes of activity that are defined by standards of excellence and have internal values that enable participants to achieve goods that could not have been achieved by any other activity. My view of values is thus that they are connected to practices (and vice versa), and that normative reasons to do, feel, and think in specific ways are given to us mainly through participation in social practices. They are given to us by our "practical reality" (Dancy, 2000).

If practices are the source of normativity, then they simultaneously work as the background that makes possible the existence of mental life. This argument can be almost formally summarized as follows: If mental life is lived in the space of reasons, and if the space of reasons is embedded in social practices, then mental life can only be lived on the background of practices. As Dewey would say: Unlike a physical response (like a reflex, which can be explained in the space of causation), a *mental* act "involves response to a thing in its *meaning*" (Dewey, 1916:29), and meanings are normative. I have followed Wittgenstein and the pragmatists, who tried to teach us that we can only think coherently about meaning if we think of it as answerable to public criteria in social practices. The meaning of a word, a gesture, or an action, is not anything private in a subjective, inner realm, but is a use in shared social practice: "The meaning of anything in the domain of human consciousness is revealed by asking what role it plays in some human practice" (Harré, 2004:6). Thus, we should "refocus the search for *principles of order* from causality and causal mechanisms to conventions, customs, habits and practices [...], and every practice is, in various ways, subject to normative appraisal" (p. 6).

The argument here seems to lead to moral relativism, for if moral normativity is embedded in social practices, and if practices differ across cultures and historical epochs, then morality apparently becomes relative to specific cultures and historical epochs. There is something true in this, for many normative conceptions are indeed relative to specific times and places. It would be foolish to deny this. But the interpretive-pragmatic account differs from social constructionist relativism in its insistence that not all normativity can be conventional. Not all kinds of moral normativity are relative to practices, for there are certain objective moral values (e.g., truthfulness, justice, and respect for rituals) that are presupposed in our having discursive practices at all (this I argued in Chap. 6). Without such values, humans could not communicate or interact the way we do. Just as the brain's synaptic functioning is the basic fabric of neurology, so the existence of such objective and nonconstructed values make up the basic fabric of psychology. Thus, without some form of social order as a *moral* order, there cannot be fully developed mental phenomena, as we know them. This point also goes back to Aristotle who argued (in MacIntyre's rendition) that "Separated from the *polis* [the social-moral order, SB], what could have been a human being becomes instead a wild animal" (MacIntyre, 1988:98). The *polis* in this sense is the only context in which a form of life is possible that is susceptible to moral reasons for action.

In Chap. 6 I argued that there is no way to account satisfactorily for human perception, action, emotion, and identity without presupposing the existence of real moral values, or – to state some alternative conceptualizations – a space of reasons (McDowell), a moral ontology (Taylor), a moral order (Harré), or a moral ecology (Brinkmann, 2004b). Human psychological functioning presupposes an order of moral values that is not of the agent's own making. Mental life is related to *reasons* (for action, feeling, thinking), and the distinction between good and bad reasons is not to be drawn in terms of subjective states of the agent, but rather in terms of objective moral values (hence the moral realism that I have defended). In a discussion of reasons and causes in psychology, Robinson has argued for irreconcilable differences between them and claimed about the former that "Life is literally pointless without them, but psychology has not the slightest notion of what to do with them" (Robinson, 1984:163). I hope to have provided at least a gist of what to do with reasons in psychology, viz. to view them as embedded in practices, as providing the basic fabric of psychological phenomena, and as linked to objective moral values (cf. Brinkmann, 2006b). My argument has been that all psychological functioning presupposes a normative order that separates adequate from inadequate functioning, and Aristotle's conception of the *virtues*, as those properties of agents that enable them to achieve practice-internal goods, was put forth as a central concept in an interpretive-pragmatic psychology. Concepts of virtue are at once descriptive and evaluative and are used to describe the central aspects of persons. I have argued that we cannot completely separate facts from values in psychological studies, for what is interesting about mental life is describable only with a vocabulary that is at once descriptive and evaluative: thick ethical concepts (e.g. concepts of virtue). These pick out real features of the world, and are indispensable when we think about what to do, evaluate our own conduct and that of others, and try to determine what kind of person we should strive to become. In short, thick ethical concepts provide us with reasons for action (Williams, 1985:140).

In the first chapter, I defended the primacy of what Dewey called "the ordinary qualitative world" as an alternative to a priori theorizing. Since this book is largely based on theory, I can be said to contradict myself on this point, but I would like to think of the interpretive-pragmatic theory (that I have consistently referred to as a *view* rather than a theory) as being "designed precisely to protect practice against unwarranted theoretical incursions," as Dunne (1997:160) says of Aristotle's practical works. My "theory" claims (in simplified terms) that there is already morality in the world – that we do not need to import it by way of theories. Of course, theories can be useful, but I grant phenomenological description a form of primacy. Life is lived before theorized about (and this also goes for moral life), although our theorizing can alter how we live (as argued most directly in Chap. 4 when I addressed "the looping effect of human kinds").

I have chosen the name "interpretive-pragmatic" for the view proposed in this book, not because I am particularly wedded to this hyphenation, but because it has been useful for me to have a quick way of pointing to my own position. Interpretive-pragmatic moral science aims to make the inevitable connections between morality and psychology visible by highlighting at once the ways in which scientific accounts of humans can assist in "making up people" (as discussed in part I) and the ways in which humans are irreducibly moral creatures, i.e., creatures that give and expect more or less good and adequate reasons for what they do, think and feel (as discussed in part II). People and the normative reasons they have for doing things are not floating in a void, but are always parts of historically developed interpretive traditions (of this *interpretive* in my hyphenation), which have, however, the possibility of being improved through the process Taylor calls "reasoning in transitions" (i.e., reasoning from within human practices). Whether something can be allowed to count as "improvement" is ultimately determined by its effects in practice (of this *pragmatic* in my hyphenation).

Challenges

I believe that a primary strength of the interpretive-pragmatic framework is that it respects our moral experience. But like any framework, it has its own weaknesses, and it can definitely be challenged, and in what follows I list a number of important challenges that I have not raised earlier, and I give short accounts of how I believe they should be answered.

- Where is the body in my account? Have I not portrayed moral agents as free-floating intellects in a sea of moral reasons?

  The interpretive-pragmatic framework should be able to articulate a better account of the bodily skills and habits that give flesh and bones to moral life than I have done here. We are not pure observing intellects that disinterestedly grasp the normative reasons of social practices that structure mental life. Rather, we literally *in*-habit the world by virtue of our bodily based habits, and a concept of bodily skill or skillful comportment is needed in order to grasp this "material" side of moral and mental life. As Harré has said: "In acting skillfully a person picks out relevant aspects of his or her external environment. What that person then does is subject to standards of correctness and propriety. A person acquires skills by training and practice. To a skilled person correct action is 'second nature'" (Harré, 2002:152). To Harré, the concept of skill is what connects the body and the material world with the normative structure of reality[2]. Although I have not given our bodily nature an extensive treatment in this book, I do not think that doing so would contradict or render invalid the interpretive-pragmatic view that I have tried to develop. The

[2] I hope it has become clear that I do not like to talk about the "material world" and the "social world" (or normative/intentional world) as two different worlds, but I have retained this way of talking simply as a shorthand. In reality, there is just one world with many different properties, some of which we call "material" and others "normative" and "social," but these are *analytic categories* that enable us to understand different kinds of processes (e.g., the differences between human kinds and natural kinds) rather than ontological distinctions.

interpretive-pragmatic account of mental and moral life certainly needs to include an account of bodily, habitual being-in-the-world, and it would be interesting to pursue this matter further, e.g., by examining the resources in the philosophies and psychologies of the body and habits found in the works of Dewey and Merleau-Ponty, for example.

- If the subject matter of psychology is mental life as "lived in the space of reasons," does this not exclude animals and small children from psychology's domain?

  This is a reasonable objection, since psychology should not just address the lives of adult human beings who can give reasons and expect such reasons in return from others, but also the lives of children who are not yet able to engage in reason giving practices and who may not have mastered language at all. Similarly, psychology has traditionally studied animal behavior, and to some it would seem to be an unacceptable loss if animals were excluded from the domain of psychology. I have no particular wish to keep non-human animals in psychology, but I do not think that my arguments about the normativity of the mental automatically disqualifies animals (or pre-linguistic children) from psychology's subject matter. The crux of the matter concerns whether animals and children can act for reasons – thus being susceptible to normativity. Recently, the Wittgensteinian scholar Hans-Johann Glock has argued that animals can indeed act for reasons. Only on a false – subjectivist, psychologistic – construal of "reasons" are animals (and small children) outside the normative sphere of reasons. If reasons, as subjectivists following Hume believe, were psychological or mental states of the agent, then it would indeed be difficult to talk about animals acting for reasons, for this would seem to presuppose a linguistic conceptualization of the relevant subjective beliefs, which animals and small children are not capable of.

  If, on the other hand, reasons are construed – and rightly so, in my opinion – as *facts*, then there does not seem to be a problem in ascribing the capacity for acting for reasons to animals and small children: "reasons for actions are facts," says Glock, "and animals are certainly capable of acting in the light of facts" (Glock, 2009:233). Most psychologists psychologize reasons and see them as subjective mental states, but if reasons are what we invoke when asked why we do what we do, then it is clear that reasons are not in general subjective mental states. My reason for taking an umbrella is that it is raining, not that I believe that it is raining. It is the weather – the facts or the states of affairs – rather than my mental state that explains why taking an umbrella is a good thing or a bad thing (p. 241). In the discussions in Chaps. 6 and 7, I likewise argued that a moral action is not moral, because it is done for a reason that refers to the agent's subjectivity (we do not, for example, count an action moral if it was done for the sake of displaying a virtue), but rather because of moral facts, or moral states of affairs that refer to other people and social situations.

  Approaching reasons as facts rather than subjective mental states solves some of the problems here. There is a difference between being able to act in the light of reasons and being able to reflect on one's having these reasons, and animals and small children may not be able to do the latter, but they are capable of doing the former (Glock, 2009:233). And, of course, we still need to investigate how creatures develop the latter capabilities from the former, but this is not a task to be undertaken in the present book. I would like to mention, however, that the best account we have of this process is in my view found in Vygotsky's developmental theory. According to Vygotsky, a child's higher mental functions (not least its abilities to think, reflect, and act for reasons) are formed when adults interpret, and act upon, the child's primitive behaviors, thereby transforming biological dispositions into social acts. Through this process, the child is gradually moved into the space of reasons – or, in other words, from first-nature to second-nature. The famous example discussed by Vygotsky (1978) concerns what happens when a child is trying to reach something by performing a grasping movement. Adults subsequently bring that something to the child, who thereby learns to perform a pointing gesture. Learning to use social signs, such as intentionally pointing one's finger, means developing a second nature and entering the space of reasons. But what is significantly *not* discussed by Vygotsky, is the moral order surrounding this development, for – to make the point simple – not everything that the child points to will be brought to the child, and there is definitely such a thing as inappropriate pointing (e.g., unsuitably pointing at someone), when the child is likely to be reprimanded rather than rewarded for pointing. Thus, an important aspect of learning to point involves learning about *when* to point, and at *what*. There is a moral normativity to pointing as to all other uses of signs in social situations and to all higher mental functions as such. Being an acting subject with the capacity for full-blown mental life, including the capacity to reflect on the reason one may have, means to be accountable in relation to this moral normativity, but a full treatment of this will have to wait for another occasion.

- Is my account not purely utopian? Is it not based on the premise that we can all become well-functioning moral agents that are able to act for good moral reasons? What about power relations that exclude certain people from moral action?

  Future work should address what *hinders* the development of our moral powers, our skills of moral perception, action, emotion, and judgment. We need to understand more about the ways in which human beings are *positioned* differently and have access to different resources in the moral domain. Not every person can verbalize, reflect, and make explicit the moral values that are in play in that person's life. This challenge is related to the first one mentioned, for a way to balance the relationship between habitual action and reflective, verbalized discussions of morality is to give the body a more central position, including the power relations that determine how habits are formed and bodies are disciplined. My perspective can perhaps be accused of favoring masculine, middle-class values in emphasizing the human capacity for making explicit the moral values we live by, whereas other forms of life put less weight on verbalizations and reflective judgments. In defense, I would say that it is a matter of careful empirical studies to give concrete contents to the moral worlds that people inhabit. I would reject the idea that a moral world, a moral order, or an evaluative background can remain *totally* implicit and impossible to articulate, for in that case, it could not have genuinely normative or intentional properties. It would thus literally be meaningless.

- Is not mental illness a psychological phenomenon, yet something that calls for causal explanation? And does this not contradict my idea that psychological phenomena are situated in normative orders?

  There are two answers to this question: The first is to emphasize that although psychological phenomena exist in normative moral orders (the space of reasons), there is no warrant for the conclusion that *all* mental life must *always* be based on reasons. The latter idea is obviously false. Most of what we do is not based on articulated reasons, but on habits (but, as Aristotle and Dewey argued, habits can be based on reasons). However, if there are no meanings whatsoever to articulate in relation to some process (say, some process in the brain), i.e., if it defies normative evaluation and is totally immune to improvement in the light of normative reasons, then the process is not a psychological process (cf. the previous objection). It is the *possibility* that something can enter the space of reasons, rather than its actual being evaluated or justified there, which determines whether something counts as a psychological phenomenon.

  The second reply is to ask: In so far as some "mental illness" is purely caused and not based on reasons, why would we consider it a suitable subject matter for psychology[3]? In a series of well-known and provocative texts, Szasz (1961, 1994) has argued that mental illness is a myth, since the concept has no literal application. Even the fact that we can "see" mental illness in the brain does not support the fact that mental illness exists, quite the contrary:

> People believe that finding brain lesions in some mental patients would prove, or already has proven, that mental illnesses exist and are 'like other illnesses'. This is an error. If mental illnesses are diseases of the central nervous system [...] then they are diseases of the brain, not the mind; if they are names of misbehaviors, then they are not diseases (Szasz, 1994:35).

Szasz argues that conceptual analysis demonstrates that the concept of illness is logically attached to physical illness, i.e., illness in tissue. Let us take depression as an example. If some depression is caused (e.g., by a brain tumor) then it is an illness of the brain and not a psychological phenomenon (although its consequences may be psychologically relevant). If, on the other hand, a person has *reasons* for feeling depressed (say, if his wife has left him), then we are in the realm of normativity, and here Szasz finds that the concept of illness has no application. Instead, Szasz prefers to talk about "problems in living." Problems in living are what psychotherapists, unlike neurologists, help their clients deal with. In the terms used in this book, the argument thus states that if there are *reasons* for some problem, then it is not an illness, and if there are only *causes*,

[3] Of course, its *consequences* are no doubt a suitable subject matter for psychology, like the psychological consequences of living with a brain tumor, which, in itself, is not something psychological. Likewise, it is relevant to investigate the psychological consequences of a natural disaster, but we do not think of the natural disaster itself as a psychological phenomenon.

then it is not a "mental" problem[4]! Hollis's (1977) argument, that reasons are enough to explain (moral) actions, is relevant here. If a person does something and we are provided with a reason that satisfactorily explains the action, then the search for explanation normally stops. Only *irrational* actions call for causal explanations, i.e., if we cannot find a *reason*-able explanation as to why someone did something. I would add that also *a-rational* behaviors, i.e., processes that can neither be described as rational or irrational (like the reflex movement of my leg as described in the introduction), call for causal explanation, since these are outside the space of reasons. In so far as something is *a-rational* in this sense, it does not belong to psychology, and in so far as it is *irrational*, it is an open question whether it does or not. In an interesting paper on psychiatric disorders, Matthews (2004) claims that "making sense of the disorder crucially involves causal explanation in terms of the factors leading to that neurological breakdown. But even then it remains true that what makes this a *disorder* consists in the alteration of the normal structuring of that kind of experience, and not in what makes that alteration come about" (p. 196). Matthews (who follows Merleau-Ponty) seeks a middle ground between explanation in terms of reasons and causes, which leads me to the next question.

- Why is it necessary to distinguish sharply between reasons and causes and claim that psychology operates in the field of reasons? Why not simply acknowledge that reasons can also be causes?

  I have avoided dealing explicitly with the difficult philosophical question whether reasons can also be causes, i.e., whether the reason we have for doing something in fact is what causes us to do it. This question involves extremely tricky philosophical problems that I cannot address in detail here, so I shall merely briefly explain why I believe that reasons and causes are so different from each other that it seems unhelpful to say that reasons can be causes. As Wittgenstein's exegetes Baker and Hacker have said: "reasons and motives are no more causes of action than the premises of a syllogism are the causes of its conclusion" (Baker & Hacker, 1982:239). Reasons-talk in general is quite different from causes-talk, for we can have a reason (to do something or believe something) without this causing us to do anything (cf. the problem of weakness of the will: when I have a reason to do something, but cannot bring myself to do it). Furthermore, unlike causes, reasons are not transitive. That is, if A is the cause of B, and B is the cause of C, then A is the cause of C (Hollis, 1977:108). But this does not go for reasons, for if A is the reason for my action B, then I am responsible for B, but I am not similarly responsible "for what others do

[4] Confer the question whether there are reasons for someone's depression or whether there are only causes. Questions of this type can involve a (moral, political) struggle as to how processes and events are to be classified; are they natural or human kinds, for example? In the current cultural climate, it may for example be easier to obtain funding for one's research into depression if the syndrome is framed in "the space of causation," but, if my arguments are valid, this may be at the expense of carrying out genuine *psychological* research about it.

autonomously because of what I set in motion" (p. 108). Responsibility and other normative concepts are not transitive in a simple way like causality. Finally, causes-explanations work by bringing particular observations under a general law, but reasons-explanations work differently, viz. by explaining "the particular by the particular" (p. 108). In general people do not act because their actions are instances of a general causal law (e.g., I do not love my wife because there is a general law specifying that humans of type X are attracted to humans of type Y, but because she is lovable!). Even if there *is* a general law, this is not the reason why we act as we do.

- Is it not illegitimate to talk about "psychology" as an agent that does certain things?

  It is true that I have sometimes talked about the rise of "psychology" as if it were a single thing, and I have analyzed certain social processes as being affected by this "thing": "psychology." The easiest answer to the objection that this seems to be an unjustified reification of a discipline is to say that, of course, it is always *people* who do things, notably psychologists who incarnate the discipline, but also numerous other people, e.g., educationalists, managers, and parents, who have come to act on the background of the psychological social imaginary. There is some truth in this, but I believe that the matter is more complex. For I would reject the view that humans have interests and intentions in abstraction from the practices and vocabularies in which they frame their interests and intentions. The psychological social imaginary does not just exist in people's heads, but in numerous practices, sites, technologies, and discourses. Our very interests and ways of acting today are in many ways *constituted* by psychological modes of understanding. Latour (2005) argues that "action" must be understood as radically dispersed, and that an actor can only act given a host of mediating factors that can usefully be framed as a network. I do not agree with all of Latour's conclusions, but I do think there is a point in saying that technologies in a broad sense (including what I called "intellectual technologies") are not just means that we employ in service of predetermined ends, for technologies determine what can intelligibly be counted as an end (I can only want to write this book because of the existence of books, disciplines, alphabets, institutions, publishers etc.). From this point of view, I hope it appears reasonable to talk about "psychology" as an agent that does things in the world, and, through its operations, enables *us* to do certain things (and restrain us from others). I have perhaps been too lax in my use of the notion "psychology," for sometimes I use it to name the discipline, and sometimes to name the subject matter[5], but, if nothing else, this at least underlines the idea that the two influence each other.

- Is it really true, as I have claimed, that psychological theories are valid to the extent that they enrich the human world of social practices? Does this go for all psychological theories, or are there other criteria of validity?

[5] In an earlier version of the text, I used "Psychology" to refer to the discipline and "psychology" to refer to the subject matter, but it appeared quite clumsy.

People who work with neuropsychology, visual perception, behavioral genetics, or psychopathology will perhaps react to my point of view and say: well, *our* theories are valid if they correspond to what the world is like independently of our theories, so validity is quite independent of the "use value" of our theories. To some extent, I accept this. We may here follow Dewey, who distinguished between two psychologies, and argued that in psychology, "all phenomena can be divided into the physiological and the social, and that when we have relegated elementary sensation and appetite to the former head, all that is left of our mental life, our beliefs, ideas and desires, falls within the scope of social psychology" (Dewey, 1917:54). The point that theories are valid to the extent that they enrich our practices goes for what Dewey calls "social psychology" only. However, my argument has also been that "social psychology" (in Dewey's sense) enjoys a certain primacy, because it is concerned with the domain where phenomena are meaningful and normative, and – as I have tried to argue – *bona fide* psychological. But rather than rejecting "physiological psychology" (again, in Dewey's sense), I would see this branch of psychology as a useful discipline that helps us throw light on psychological phenomena, but *which in itself does not deal with psychological phenomena*. The brain, the retina, the genes, or a dysfunctional serotonin level are not psychological objects, but no doubt *relevant* to the study of psychological objects, which, to repeat myself one more time, are irreducibly normative *qua* psychological. And here, validity is a practical and moral issue.

Psychology's Vocabulary Reconsidered

Psychologists may read this book and say: "Very well, but so what? What are the consequences for how to do psychology?" It has not been my intention to formulate detailed guidelines for psychologists as "moral scientists," but rather to examine psychology as a moral science. Nonetheless, I hope the "so what reaction" can be moderated if we reconceptualize some central concepts in the science of psychology in the light of moral normativity. I shall not do so in any great detail, but merely provide some hints at how the field and its main concepts can be reordered if my theses are valid and if it is true that psychological phenomena are moral phenomena.

Person/Self

From the interpretive-pragmatic point of view, the category "person" is not value-neutral. To be a person is to be able to act intentionally and, to some extent, to be responsible for what one does. To be a person is to be situated in a normative framework (the space of reasons) that is constitutive of actions as intentional and of humans as persons. This normative framework is only possible within social practices where there are standards of what counts as a reason for what. Thus a *person* can be understood as a "location" of responsibility and accountability in a normative framework (Harré, 1983). What about the concept of *self*? Is this not value neutral? No, I take the concept of self to be largely synonymous with the concept of "identity"[6]. As I argued in Chap. 6, persons have selves/identities to the extent that they have a sense of how their biographies are connected to larger moral issues to which they are committed. A self or an identity is thus formed by moral commitments, or, more radically, *is* the person's concrete moral commitments. A quote from one of John Dewey's letters (to Thomas Davidson, who directed a summer school of philosophy in the US) articulates well how our selves or identities are bound up with moral values and commitments:

> But when you say that I have no answer to the question, "Why am I *bound*?" I rise up in protest. Why, what am *I*? *I am* nothing but this binding; it is my bindings (in conduct) which make me what I am. "Why am I bound to do good?" Because that is what *I am* (quoted from Martin, 2002:123).

Dewey argued that when we use the little everyday word "I," we "accept and affirm a responsibility" (Dewey, 1925:233). And Wittgenstein argued later that when we express our intentions we are not describing some private, mental realm or making predictions, but we are publicly *committing ourselves* to fulfilling something (Harré & Tissaw, 2005:233), and thus place us deliberately in the space of reasons.

Personality

The study of personality is a major issue in psychology and a dominant approach is trait psychology, which conceives of personality as made up of more or less stable traits. What is often not discussed, however, is that when we talk about traits or characteristics that form someone's personality, we talk about things with clear moral implications. It is almost too obvious to mention, but if we take the most widely used model of personality in contemporary psychology – Costa and McCrae's model of the allegedly robust Big Five traits (McCrae & Costa, 1987) – then its five factors of personality (neuroticism, extraversion, openness, agreeableness and conscientiousness) are all clearly morally laden. In fact, they are all what I have referred to as "thick ethical concepts," i.e., concepts that pick out moral features of the world, which cannot be picked out by morally neutral concepts. Studying personality from the interpretive-pragmatic point of view would demand a fuller sensitivity to the moral implications of these concepts and would try to relate them to work done in virtue theory. It would also imply consideration of the human activities relative to which such personality and trait terms have moral meaning (for although conscientiousness is a virtue, a conscientious torturer is in fact less attractive than a conscientious medical doctor!).

[6]Cf. Taylor's *Sources of the Self*, which bears the subtitle *The Making of the Modern Identity*.

Cognition

Emmanuel Levinas has characterized the modern naturalistic outlook on the world as one that "places subject and object in the same world, which it calls nature, and studies their relation as a relation of causality" (Levinas, 1973:15). Most psychological studies of human cognition share this outlook. Cognition is studied as a causal relationship in which the cognizer is causally affected by the world and in turn causally affects the world. The interpretive-pragmatic view would simply reject the view that cognition is a pure causal process by pointing out that cognition in the sense of *knowledge* is an irreducible normative notion. In his 1956 classic, *Empiricism and the Philosophy of Mind* (1997), Wilfrid Sellars argued that when we talk about an episode as one of knowing, we inevitably raise the issue of its normative status. In order for something to count as knowledge, we have to be able to justify what we think we know. That a belief is true is not enough for it to count as knowledge, for it may (for example) be the result of a lucky guess. As Sellars said: "The essential point is that in characterizing an episode or a state as that of *knowing*, we are not giving an empirical description of that episode or state; we are placing it in the logical space of reasons, of justifying and being able to justify what one says" (p. 76). An interpretive-pragmatic approach to cognition would see all cognition terms as normative success terms, and all cognition as basically a form of problem-solving (this was also Dewey's position).

Perception

Any account of perception must operate with a normative difference between veridical and non-veridical perception. To perceive something is to *succeed* in some way, viz. to perceive it *accurately*, and the inability to perceive can, at least sometimes, be seen as a moral defect (if, for example, someone repeatedly overlooks the needs of other people). I have argued (Chap. 6) that perception is a value-realizing activity in the sense that it is not primarily a passive receptivity of the world's stimuli (which can be explained in causal and value neutral terms), but consists of someone's active search for the affordances or *values* of objects, persons, and situations. The moral and even spiritual nature of perception has also been commented upon by Nietzsche (1889:75): "Learning to see – habituating the eye to repose, to patience, to letting things come to it; learning to defer judgment, to investigate and comprehend the individual case in all its aspects – This is the *first* preliminary schooling in spirituality." I have in several places put emphasis on the perception of moral states of affair, especially in Chap. 6 (where I referred to the Gestalt, ecological, and phenomenogical schools).

Emotion

According to the interpretive-pragmatic framework, emotions are not morally neutral psychological happenings, for, just like other mental events, emotions are what they are, only when given some evaluative background. In the Aristotelian tradition, emotions can help the agent orient herself adequately in moral space. To have emotions is a way of knowing the value-laden features of the world. Thus my sadness can be a morally adequate response to a situation that normatively calls for this kind of reaction. Feeling sad is a way of knowing a sad situation. We can thus feel rightly and wrongly, just as we can act rightly and wrongly. It is a fact that we are sometimes morally praised and blamed for our emotions, and not just for our actions, and, as Baerveldt and Voestermans (2005) have recently argued, emotions are our primary way of being tied to the normative structure of reality.

Action

In Chap. 6, I argued that action is a moral concept in the sense that actions (in contrast to behaviors) are based on reasons, and reasons can be good or bad, more or less adequate. We distinguish actions in light of the reasons there are for doing specific things, and the same physical pattern of behavior (e.g., mowing the lawn) can represent different actions in accordance with the reasons on which the given action is based (cf. the difference between mowing the lawn in order to please one's wife vs. in order to take exercise). Such reasons are relative to social practices. We can only intend to do certain things as participants in structured activities or practices. In line with Charles Taylor, the interpretive-pragmatic framework denies that there could be human actions as we know them if actors could only evaluate what they do in the light of their immediate preferences and desires. In addition there has to be moral values that stand or fall independently of the actor's preferences, and these thus deserve to be called "objective moral values." As argued by Louch: "observation, description and explanation of human action is only possible by means of moral categories" (Louch, 1966:21), for "To identify a piece of behavior as an action is already to describe experience by means of moral concepts"(pp. 26–27). We cannot pick out an event as an action without having some evaluative standpoint, often articulated with thick ethical concepts.

Memory

Like perception, memory belongs to the class of achievement concepts. To remember something is to do so correctly, which means that memory is normative. Furthermore, as Harré (2002:164) has argued, when we remember something in everyday life, we very rarely have access to forensic evidence of what happened. Instead we certify our memory through negotiation with others, and the tendency of scientific psychology to take people's performances in laboratory situations as evidence of how well they remember *tout court* is a choice with certain (lamentable) moral effects. In laboratory situations, to mention just one example, younger people generally do better than older people, but when young and old people are taking part in conversations, they are equally good (p. 164). Thus, Harré concludes that it is not only false but also *immoral* to state that older people do not remember as well as younger people, for such statements "leak out into the lay world, affecting the attitudes of the social services and employers to the capacities of older people" (p. 164). This is an example of the moral consequences of the looping effect of human kinds.

Development

In this book, I have said little about psychological development in general, or moral development in particular, but the interpretive-pragmatic view would agree with Jerome Bruner that "theories of human development constitute a *policy science*, a science whose intrinsic object is not simply to describe but to prescribe alternative optimal ways of achieving certain outcomes" (Bruner, 1986:20; see also Smith, 2006). When researchers study human development they cannot simply be engaged in value neutral description, for they always choose to hold certain patterns up as "normal" or "natural," while other patterns are seen as deviant. Such choices are deeply normative, and naturally affect the ways children and adults are conceived and acted upon in institutions, schools, and families. For example, the discovery that small children are much more "competent" than previously thought has in significant ways altered the attitudes of parents and educators towards these children, and they now grow up to become different persons (more independent, reflective, verbalizing) than in previous times when children were seen as much more passive and vulnerable. Developmental psychology should recognize its own looping effects, and, in Bruner's words, see itself as "a device for describing how to navigate in the value-laden and constructed world of symbols and technology created by the society for its own regulation" (p. 21).

Bruner's remark could stand for psychology as a whole, according to the present study: Psychology should be conceived as a device that describes, regulates, and to some extent constitutes, ways that humans navigate in a value-laden world. I have described the device of psychology as having moral *effects* in being able to change how persons think about the oughtness of human life. At the same time I have argued that persons themselves are moral creatures. Thus, if acting, feeling, and thinking persons are the subject matter of psychology, then this science has a subject matter with irreducible moral *content*.

References

Annas, J. (2001). Ethics and morality. In L. C. Becker & C. B. Becker (Eds.), *Encyclopedia of ethics* (2nd ed.). London: Routledge.

Arendt, H. (1958). *The human condition*. Chicago: University of Chicago Press (This edition 1998).

Aristotle (1976). *Nichomachean ethics*. London: Penguin

Averill, J. R. (1982). *Anger and aggression: An essay on emotion*. New York: Springer.

Backe, A. (2001). John Dewey and early Chicago functionalism. *History of Psychology, 4*, 323–340.

Baerveldt, C., & Voestermans, P. P. (2005). Culture, emotion and the normative structure of reality. *Theory and Psychology, 15*, 449–473.

Baker, G. P., & Hacker, P. M. S. (1982). The grammar of psychology: Wittgenstein's Bemerkungen Über die Philosophie der Psychologie. *Language and Communication, 2*, 227–244.

Baldwin, J. M. (1901). *Dictionary of philosophy and psychology*. London: MacMillan.

Barash, D. (1982). *Sociobiology and behavior*. New York: Elsevier.

Barash, D. (2000). Evolutionary existentialism, sociobiology, and the meaning of life. *Bioscience, 50*, 1012–1017.

Bauman, Z. (1996). From pilgrim to tourist – or a short history of identity. In S. Hall & P. du Gay (Eds.), *Questions of cultural identity*. London: Sage.

Bauman, Z. (2007). *Consuming life*. Cambridge: Polity Press.

Bellah, R. N., Madsen, R., Sullivan, W. M. S. A., & Tipton, S. M. (1985). *Habits of the heart: Individualism and commitment in American life*. Berkeley: University of California Press.

Bellah, R. N. (1983). The ethical aims of social inquiry. In N. Haan, R. N. Bellah, P. Rabinow, & W. M. Sullivan (Eds.), *Social science as moral inquiry*. New York: Columbia University Press.

Bem, S., & Looren de Jong, H. (1997). *Theoretical issues in psychology*. London: Sage.

Berlin, I. (1999). *The roots of romanticism*. London: Chatto and Windus.

Bertelsen, P. (2002). Will and self in augustine: Double-bind and taboo. *Theory and Psychology, 12*, 749–776.

Blum, L. (1991). Moral perception and particularity. *Ethics, 101*, 701–725.

Boltanski, L., & Chiapello, E. (2005). *The new spirit of capitalism*. London: Verso.

Bourdieu, P. (1977). *Outline of a theory of practice*. Cambridge: Cambridge University Press.

Bourdieu, P. (1998). *Practical reason: On the theory of action*. Cambridge: Polity Press.

Bredo, E. (1998). Evolution, psychology, and John Dewey's critique of the reflex arc concept. *The Elementary School Journal, 98*, 447–466.

Brinkmann, S. (2004a). Psychology as a moral science: Aspects of John Dewey's psychology. *History of the Human Sciences, 17*, 1–28.

Brinkmann, S. (2004b). The topography of moral ecology. *Theory and Psychology, 14*, 57–80.

Brinkmann, S. (2005). Psychology's facts and values: A perennial entanglement. *Philosophical Psychology, 18*, 749–765.

Brinkmann, S. (2006a). Damasio on mind and emotions: A conceptual critique. *Nordic Psychology, 58*, 366–380.
Brinkmann, S. (2006b). Mental life in the space of reasons. *Journal for the Theory of Social Behaviour, 36*, 1–16.
Brinkmann, S. (2006c). Questioning constructionism: Toward an ethics of finitude. *Journal of Humanistic Psychology, 46*, 92–111.
Brinkmann, S. (2008). Identity as self-interpretation. *Theory and Psychology, 18*, 404–422.
Brinkmann, S. (2009). Guilt in a fluid culture: A view from positioning theory. *Culture and Psychology., 16*(2), 253–266.
Bruner, J. (1986). Value presuppositions of developmental theory. In L. Cirillo & S. Wapner (Eds.), *Value presuppositions in theories of human development*. Hillsdale, NJ: Lawrence Erlbaum Associates.
Bruner, J. (1999). *Mening i handling [Acts of meaning]*. Århus: Forlaget Klim.
Burkitt, I. (2002). Technologies of the self: Habitus and capacities. *Journal for the Theory of Social Behaviour, 32*, 219–237.
Burr, V. (1995). *An Introduction to social constructionism*. London: Routledge.
Buss, D. M. (1999). *Evolutionary psychology*. Boston: Allyn and Bacon.
Caporael, L. R. (2001). Evolutionary psychology: Toward a unifying theory and a hybrid science. *Annual Review of Psychology, 52*, 607–628.
Casebeer, W. D. (2003). *Natural ethical facts: Evolution, connectionism, and moral cognition*. Cambridge, MA: MIT.
Cavell, S. (1969). *Must we mean what we say?* New York: Charles Scribner's Sons.
Charland, L. C. (2004). Character: Moral treatment and the personality disorders. In J. Radden (Ed.), *The philosophy of psychiatry: A companion*. Oxford: Oxford University Press.
Christians, C. G. (2000). Ethics and politics in qualitative inquiry. In N. K. Denzin & Y. S. Lincoln (Eds.), *Handbook of qualitative research*. London: Sage.
Colby, A., & Damon, W. (1992). *Some do care: Contemporary lives of moral commitment*. New York: The Free Press.
Coles, R. (1992). *Self/power/other: Political theory and dialogical ethics*. Ithaca: Cornell University Press.
Collin, F. (1990). Naturlige klasser, semantik og metode i samfundsvidenskaberne. *Filosofiske studier, 11*, 7–24.
Costall, A. (2004a). From Darwin to Watson (*and* Cognitivism) and back again: The principle of animal-environment mutuality. *Behavior and Philosophy, 32*, 179–195.
Costall, A. (2004b). From direct perception to the primacy of action: A closer look at James Gibson's ecological approach to psychology. In G. Bremmer & A. Slater (Eds.), *Theories of infant development*. Oxford: Blackwell.
Craighead, W. E., & Nemeroff, C. B. (2001). *The corsini encyclopedia of psychology and behavioral science*. New York: Wiley.
Cushman, P., & Gilford, P. (1999). From emptiness to multiplicity: The self at the year 2000. *The Psychohistory Review, 27*, 15–31.
Cushman, P. (1990). Why the self is empty: Toward a historically situated psychology. *The American Psychologist, 45*, 599–611.
Cushman, P. (1995). *Constructing the self, constructing America: A cultural history of psychotherapy*. Cambridge, MA: Perseus Publishing.
Damasio, A. (2003). *Looking for spinoza: Joy, sorrow, and the feeling brain*. Orlando: Harcourt.
Dancy, J. (2000). *Practical reality*. Oxford: Oxford University Press.
Danziger, K. (1990). *Constructing the subject: Historical origins of psychological research*. Cambridge: Cambridge University Press.
Danziger, K. (1996). The practice of psychological discourse. In C. F. Graumann & K. J. Gergen (Eds.), *Historical dimensions of psychological discourse*. Cambridge: Cambridge University Press.
Danziger, K. (1997a). *Naming the mind: How psychology found its language*. London: Sage.
Danziger, K. (1997b). The historical formation of selves. In J. D. Ashmore & L. Jussim (Eds.), *Self and social identity: Fundamental issues*. New York: Oxford University Press.

Danziger, K. (1999). Natural kinds, human kinds, and historicity. In W. Maiers, B. Bayer, B. Duarte Esgalhado, R. Jorna, & E. Schraube (Eds.), *Challenges to theoretical psychology*. North York: Captus Press.

Danziger, K. (2003). Where history, theory, and philosophy meet: The biography of psychological objects. In D. B. Hill & M. J. Kral (Eds.), *About psychology: Essays ate the crossroads of history, theory, and philosophy*. Albany, NY: State University of New York Press.

Darley, J. M., & Schultz, T. R. (1990). Moral rules: Their content and acquisition. *Annual Review of Psychology, 41*, 556.

Davies, B., & Harré, R. (1999). Positioning and personhood. In R. Harré & L. van Langenhove (Eds.), *Positioning theory: Moral contexts of intentional action*. Oxford: Blackwell.

Davydova, I., & Sharrock, W. (2003). The rise and fall of the fact/value distinction. *The Sociological Review, 51*, 357–375.

Dawkins, R. (1986). *The blind watchmaker*. New York: W.W. Norton & Co.

De Sousa, R. (2001). Moral emotions. *Ethical Theory and Moral Practice, 4*, 109–126.

Dewey, J. (1891). Moral theory and practice. *International Journal of Ethics, 1*, 186–203.

Dewey, J. (1896). The reflex arc concept in psychology. *Psychological Review, 3*, 357–370.

Dewey, J. (1916). *Democracy and education*. New York: The Free Press. This edition published 1966.

Dewey, J. (1917). The need for social psychology. In J. A. Boydston (Ed.), *The middle works, volume 10: 1916–1917*. Carbondale: Southern Illinois University Press. This edition published 1980.

Dewey, J. (1920). *Reconstruction in philosophy*. Boston: The Beacon Press. Second enlarged edition published 1948.

Dewey, J. (1922). *Human nature and conduct: An introduction to social psychology*. New York: The Modern Library. This edition published 1930.

Dewey, J. (1925). *Experience and nature*. Chicago: Open Court.

Dewey, J. (1929). *The quest for certainty*. New York: Capricorn Books. This edition 1960.

Dewey, J. (1934). *Art as experience*. New York: Perigee Books. This edition 1980.

Dewey, J. (1976). *Lectures on psychological and political ethics: 1898*. New York: Hafner Press.

Dewey, J., & Bentley, A. (1949). *Knowing and the known*. Boston: Beacon Press. This edition 1960.

Dilthey, W. (1977). *Descriptive psychology and historical understanding*. The Hague: Martinus Nijhoff. First published 1894.

Dreyfus, H. (1991). *Being-in-the-world – A commentary on heidegger's being and time, division I*. Cambridge, MA: MIT.

Dreyfus, H., & Dreyfus, S. (1990). What is morality? A phenomenological account of the development of ethical expertise. In D. Rasmussen (Ed.), *Universalism vs. communitarianism: Contemporary debates in ethics*. Cambridge, MA: MIT.

Dreyfus, H., & Rabinow, P. (1982). *Michel foucault: Beyond structuralism and hermeneutics*. Hemel Hempstead: The Harvester Press.

Dunne, J. (1997). *Back to the rough ground*. Notre Dame, IN: University of Notre Dame Press. Paperback ed.

Eisenberg, N. (1996). Caught in a narrow kantian perception of prosocial development: Reactions to Campbell and Christopher's critique of moral development theory. *Developmental Review, 16*, 48–68.

Eisenberg, N. (2000). Emotion, regulation, and moral development. *Annual Review of Psychology, 51*, 665–697.

Erikson, E. H. (1968). *Identity: Youth and crisis*. New York: W.W. Norton.

Flack, J. C., & de Waal, F. B. M. (2000). 'Any Animal Whatever': Darwinian building blocks of morality in monkeys and apes. *Journal of Consciousness Studies, 7*, 1–29.

Flanagan, O. (1991). *Varieties of moral personality: Ethics and psychological realism*. Cambridge, MA: Harvard University Press.

Flyvbjerg, B. (1991). Sustaining non-rationalized practices: Body-mind, power and situational ethics. An interview with Hubert and Stuart Dreyfus. *Praxis International, 11*, 93–113.

Flyvbjerg, B. (2001). *Making social science matter – Why social inquiry fails and how it can succeed again*. Cambridge: Cambridge University Press.

Foot, P. (2001). *Natural goodness*. Oxford: Oxford University Press.

Foucault, M. (1977). *Discipline and punish: The birth of the prison*. New York: Pantheon.

Foucault, M. (1980). *The history of sexuality, volume 1: An introduction*. New York: Random House.

Foucault, M. (1984). On the genealogy of ethics: An overview of work in progress. In P. Rabinow (Ed.), *The foucault reader*. London: Penguin.

Foucault, M. (1988a). *Technologies of the self*. London: Tavistock Publications.

Foucault, M. (1988b). The political technology of individuals. In *Technologies of the self* London: Tavistock Publications.

Foucault, M. (1988c). Truth, power, self. In *Technologies of the self* London: Tavistock Publications.

Foucault, M. (1990). *The history of sexuality, vol. 3: The care of the self*. London: Penguin.

Foucault, M. (1993). About the beginning of the hermeneutics of the self: Two lectures at Dartmouth. *Political Theory, 21*, 198–227.

Foucault, M. (1994a). Governmentality. In J. D. Faubion (Ed.), *Power: Essential works of Michel Foucault, vol. 3*. London: Penguin.

Foucault, M. (1994b). The subject and power. In J. D. Faubion (Ed.), *Power: Essential works of Michel foucault, vol. 3*. London: Penguin.

Foucault, M. (1998a). Nietzsche, genealogy, history. In J. D. Faubion (Ed.), *Aesthetics, method, and epistemology: Essential works of foucault, volume two*. New York: The New Press.

Foucault, M. (1998b). Philosophy and psychology. In J. D. Faubion (Ed.), *Aesthetics, method, and epistemology: Essential works of foucault*. New York: The New Press.

Foucault, M. (2001). *The order of things: An archeology of the human sciences*. London: Taylor & Francis. First published 1966.

Frankfurt, H. (1988). Freedom of the will and the concept of a person. In *The importance of what we care about* Cambridge: Cambridge University Press.

Gadamer, H. G. (1960). *Truth and method*. New York: Continuum. Second revised edition published 2000.

Gantt, E. E., & Reber, J. S. (1999). Sociobiological and social constructionist accounts of altruism: A phenomenological critique. *Journal of Phenomenological Psychology, 30*, 14–38.

Gazzaniga, M., & Heatherton, T. (2003). *Psychological science: Mind, brain and behavior*. New York: Norton.

Geertz, C. (1973). *The interpretation of cultures*. New York: Basic Books.

Geertz, C. (1983). *Local knowledge: Further essays in interpretive anthropology*. New York: Basic Books.

Gergen, K. (1992). Social construction and moral action. In D. N. Robinson (Ed.), *Social science and moral judgment*. London: Academic Press.

Gergen, K. (1994). *Realities and relationships*. Cambridge, MA: Harvard University Press.

Gergen, K. (1999). *An invitation to social construction*. London: Sage.

Gergen, K. (2001a). Psychological science in a postmodern context. *The American Psychologist, 56*, 803–813.

Gergen, K. J. (1973). Social psychology as history. *Journal of Personality and Social Psychology, 26*, 309–320.

Gergen, K. J. (2001b). *Social construction in context*. London: Sage.

Gerrod Parrott, W. (2003). Positioning and the emotions. In R. Harré & F. M. Moghaddam (Eds.), *The self and others: Positioning individuals and groups in personal, political, and cultural contexts*. London: Praeger.

Gibson, J. J. (1986). *The ecological approach to visual perception*. Hillsdale, NJ: Lawrence Erlbaum Associates. First published 1979.

Giddens, A. (1976). *New rules of sociological method*. London: Hutchinson.

Gigerenzer, G. (1996). From tools to theories: Discovery in cognitive psychology. In C. F. Graumann & K. J. Gergen (Eds.), *Historical dimensions of psychological discourse*. Cambridge: Cambridge University Press.

Gilligan, C. (1982). *In a different voice*. Cambridge, MA: Harvard University Press.
Giorgi, A. (1992). An exploratory phenomenological psychological approach to the experience of the moral sense. *Journal of Phenomenological Psychology, 23*, 50–86.
Giorgi, A. (2006). Facts, vales and the psychology of the human person. *Indo-Pacific Journal of Phenomenology, 6*, 1–17.
Glock, H.-J. (2009). Can animals act for reasons? *Inquiry, 3*, 232–254.
Graumann, C. F. (1996). Psyche and her descendants. In C. F. Graumann & K. J. Gergen (Eds.), *Historical dimensions of psychological discourse*. Cambridge: Cambridge University Press.
Greer, S. (2003). Self-esteem and the demoralized self: A genealogy of self research and measurement. In D. B. Hill & M. J. Kral (Eds.), *About psychology: Essays at the crossroads of history, theory, and philosophy*. Albany, NY: SUNY Press.
Habermas, J. (1993). *Justification and application: Remarks on discourse ethics*. Cambridge: Polity Press.
Hacker, P. M. S. (2004). The conceptual framework for the investigation of emotions. *International Review of Psychiatry, 16*, 199–208.
Hacking, I. (1986). Making up people. In T. C. Heller, M. Sosna, & D. Wellbery (Eds.), *Reconstructing individualism*. Stanford: Stanford University Press.
Hacking, I. (1990). *The taming of chance*. Cambridge: Cambridge University Press.
Hacking, I. (1995a). *Rewriting the soul*. Princeton, NJ: Princeton University Press.
Hacking, I. (1995b). The looping effect of human kinds. In D. Sperber, D. Premack, & A. J. Premack (Eds.), *Causal cognition: A multidisciplinary debate*. Oxford: Clarendon Press.
Hacking, I. (1998). *Mad travelers: Reflections on the reality of transient mental disease*. Charlottesville: University Press of Virginia.
Hacking, I. (1999). *The social construction of what?* Cambridge, MA: Harvard University Press.
Hacking, I. (2001). Degeneracy, Criminal Behavior, and looping. In D. Wasserman & R. Wachbroit (Eds.), *Genetics and criminal behavior*. Cambridge: Cambridge University Press.
Hacking, I. (2004). Between Michel Foucault and Erving Goffman: Between discourse in the abstract and face-to-face interaction. *Economy and Society, 33*, 277–302.
Hall, S. (1996). Introduction: Who needs 'identity'? In S. Hall & P. du Gay (Eds.), *Questions of cultural identity*. London: Sage.
Harré, R. (1983). *Personal being*. Oxford: Basil Blackwell.
Harré, R. (2002). *Cognitive science: A philosophical introduction*. London: Sage.
Harré, R. (2004). Staking our claim for qualitative psychology as science. *Qualitative Research in Psychology, 1*, 3–14.
Harré, R., & Robinson, D. N. (1995). On the primacy of duties. *Philosophy, 70*, 513–532.
Harré, R., & Tissaw, M. (2005). *Wittgenstein and psychology: A practical guide*. Aldershot: Ashgate.
Harré, R., & van Langenhove, L. (1999). The dynamics of social episodes. In R. Harré & L. van Langenhove (Eds.), *Positioning theory*. Oxford: Blackwell.
Hartle, A. (1983). *The modern self in rousseau's confessions: A reply to St. Augustine*. Notre Dame: University of Notre Dame Press.
Hartmann, M., & Honneth, A. (2006). Paradoxes of capitalism. *Constellations, 13*, 41–58.
Hatab, L. (1995). Ethics and finitude: Heideggerian contributions to moral philosophy. *International Philosophical Quarterly, 35*, 403–417.
Hauser, M. D. (2006). *Moral minds: How nature designes our universal sense of right and wrong*. New York: HarperCollins.
Heft, H. (2001). *Ecological psychology in context: James Gibson, Roger Barker, and the Legacy of William James's Radical Empiricism*. Mahwah, NJ: Erlbaum.
Hegel, G. W. F. (1821). *Elements of the philosophy of right*. Cambridge: Cambridge University Press. This edition published 1991.
Hegel, G. W. F. (1977). *Phenomenology of spirit*. Oxford: Oxford University Press. First published 1807.
Heidegger, M. (1927). *Being and time*. New York: HarperCollins Publishers. This edition published 1962.

Held, B. S. (2004). The negative side of positive psychology. *Journal of Humanistic Psychology, 44*, 9–46.

Helwig, C. C., Tisak, M. S., & Turiel, E. (1990). Children's social reasoning in context: Reply to Gabennesch. *Child Development, 61*, 2069–2078.

Helwig, C. C., & Turiel, E. (2002). Civil liberties, autonomy, and democracy: Children's perspectives. *International Journal of Law and Psychiatry, 25*, 253–270.

Helwig, C. C., Turiel, E., & Nucci, L. P. (1996). The virtues and vices of moral developmental theories. *Developmental Review, 16*, 69–107.

Herzog, M. (1995). William James and the development of phenomenological psychology in Europe. *History of the Human Sciences, 8*, 29–46.

Hodges, B. H., & Baron, R. M. (1992). Values as constraints on affordances. *Journal for the Theory of Social Behaviour, 22*, 263–294.

Hoffman, M. (2000). *Empathy and moral development: Implications for caring and justice*. Cambridge: Cambridge University Press.

Høgh-Olesen, H. (2004). Offerets grundform – deling og socialitet hos mennesket og de øvrige primater. *Bulletin fra Forum for Antropologisk Psykologi, 14*, 6–36.

Holiday, A. (1988). *Moral powers: Normative necessity in language and history*. London: Routledge.

Hollis, M. (1977). *Models of man: Philosophical thoughts on social action*. Cambridge: Cambridge University Press.

Honneth, A. (1996). Pathologies of the social: The past and present of social philosophy. In D. M. Rasmussen (Ed.), *Handbook of critical theory*. Oxford: Blackwell.

Honneth, A. (2002). Between hermeneutics and Hegelianism: John McDowell and the challenge of moral realism. In N. H. Smith (Ed.), *Reading McDowell: On mind and world*. London: Routledge.

Honneth, A. (2004). Organized self-realization. *European Journal of Social Theory, 7*, 463–478.

Hook, D. (2003). Analogues of power: Reading psychotherapy through the sovereignty-discipline-government complex. *Theory and Psychology, 15*, 605–628.

Horwitz, A. V., & Wakefield, J. C. (2005). The age of depression. *The Public Interest, 158*, 39–58.

Huebner, A., & Garrod, A. (1991). Moral reasoning in a karmic world. *Human Development, 34*, 341–352.

Hume, D. (1978). *A treatise of human nature: Being an attempt to introduce the experimental method of reasoning into moral subjects*. Oxford: Clarendon Press. First published 1739.

Husserl, E. (1954). *Die Krisis der europäischen Wissenschaften und die tranzendentale Phänomenologie*. Haag: Martinus Nijhoff.

Illouz, E. (2007). *Cold intimacies: The making of emotional capitalism*. Cambridge: Polity Press.

Illouz, E. (2008). *Saving the modern soul*. Berkeley: University of California Press.

Jackson Lears, T. J. (1983). From salvation to self-realization: Advertising and the therapeutic roots of the consumer culture, 1880-1930. In R. W. Fox & T. J. Jackson Lears (Eds.), *The culture of consumption: Critical essays in American history* (pp. 1880–1980). New York: Pantheon Books.

James, W. (1912). *Essays in radical empiricism*. New York: Longman Green & Co.

James, W. (1950). *The principles of psychology, vol. II*. New York: Dover. First published 1890.

Jansz, J., & van Drunen, P. (2004). *A social history of psychology*. Oxford: Blackwell.

Jansz, J. (2004). Psychology and society: An overview. In J. Jansz & P. van Drunen (Eds.), *A social history of psychology*. Oxford: Blackwell.

Jarzombek, M. (2000). *The psychologizing of modernity: Art, architecture, and history*. Cambridge: Cambridge University Press.

Kant, I. (1781). *Kritik af den rene fornuft*. Frederiksberg: Det Lille Forlag. This edition published 2002.

Kazdin, A. E. (2000). *Encyclopedia of psychology*. Oxford: Oxford University Press.

Kemple, T. (2007). Spirits of late capitalism. *Theory, Culture and Society, 24*, 147–159.
Kendler, H. (1999). The role of value in the world of psychology. *The American Psychologist, 54*, 828–835.
Kendler, H. (2002). Psychology and ethics: Interactions and conflicts. *Philosophical Psychology, 15*, 289–308.
Kessen, W., & Cahan, E. D. (1986). A century of psychology: From subject to object to agent. *American Scientist, 74*, 640–649.
Koch, S. (1981). The nature and limits of psychological knowledge: Lessons of a quarter century qua science. *The American Psychologist, 36*, 257–269.
Koffka, K. (1940). Problems in the psychology of art. In R. Bernheimer, R. Carpenter, K. Koffka, & M. C. Nahm (Eds.), *Art: A Bryn Mawr symposium*. New York: Oriole Editions. This edition published 1972.
Kohlberg, L. (1971). From is to ought: How to commit the naturalistic fallacy and get away with it in the study of moral development. In T. Mischel (Ed.), *Cognitive development and epistemology*. New York: Academic Press.
Kohlberg, L. (1981). *Essays on moral development volume 1 – The philosophy of moral development*. San Fransisco: Harper and Row.
Kohlberg, L. (1984). *Essays on moral development volume 2 – The psychology of moral development*. San Francisco: Harper and Row.
Köhler, W. (1944). Value and fact. *The Journal of Philosophy, 41*, 197–212.
Köhler, W. (1959). *The place of value in a world of facts*. New York: Meridian Books. First published 1938.
Korsgaard, C. (1998). Motivation, metaphysics, and the value of the self: A Reply to Ginsborg, Guyer, and Schneewind. *Ethics, 109*, 49–66.
Kvale, S. (2003). The church, the factory and the market: Scenarios for psychology in a postmodern age. *Theory and Psychology, 13*, 579–603.
Lacey, H., & Schwartz, B. (1996). The formation and transformation of values. In W. O'Donohue & R. F. Kitchener (Eds.), *The philosophy of psychology*. London: Sage.
LaFollette, H. (2000). Pragmatic ethics. In H. LaFollette (Ed.), *The Blackwell guide to ethical theory*. Oxford: Blackwell.
Laitinen, A. (2002a). Charles Taylor and Paul Ricoeur on self-interpretations and narrative identity. In R. Huttunen, H. Heikkinen, & L. Syrjälä (Eds.), *Narrative research: Voices of teachers and philosophers*. Jyväskylä: SoPhi.
Laitinen, A. (2002b). Culturalist moral realism. In A. Laitinen & N. H. Smith (Eds.), *Perspectives on the philosophy of Charles Taylor*. Helsinki: Societas Philosophica Fennica.
Langman, L. (2003). From subject to citizen to consumer: Embodiment and the mediation of hegemony. In R. H. Brown (Ed.), *Politics of selfhood: Bodies and identities in global capitalism*. Minneapolis: University of Minnesota Press.
Lapsley, D. K. (1996). *Moral psychology*. Boulder: Westview Press.
Latour, B. (1988). *The pasteurization of France*. Cambridge, MA: Harvard University Press.
Latour, B. (2000). On the partial existence of existing and nonexisting objects. In L. Daston (Ed.), *Biographies of scientific objects* (pp. 247–269). Chicago: University of Chicago Press.
Latour, B. (2005). *Reassembling the social*. Oxford: Oxford University Press.
Lave, J., & Wenger, E. (1991). *Situated learning: Legitimate peripheral participation*. Cambridge: Cambridge University Press.
Leahey, T. (1994). *A modern history of psychology* (2nd ed.). Englewood Cliffs, NJ: Prentice-Hall.
Lefebvre, H. (1968). *Everyday life in the modern world*. London: The Penguin Press. This edition published 1971.
Levinas, E. (1973). *The theory of intuition in Husserl's phenomenology*. Evanston: Northwestern University Press.
Levine, P. (1998). *Living without philosophy: On narrative, rhetoric, and morality*. Albany, NY: State University of New York Press.

Lilleleht, E. (2003). Progress and power: Exploring the disciplinary connections between moral treatment and psychiatric rehabilitation. *Philosophy, Psychiatry and Psychology, 9*, 167–182.
Løgstrup, K. E. (1956). *The ethical demand*. Notre Dame: University of Notre Dame Press. This edition published 1997.
Louch, A. R. (1966). *Explanation and human action*. Christchurch: Cybereditions Corporation. This edition published 2000.
Lovibond, S. (1983). *Realism and imagination in ethics*. Minneapolis: University of Minnesota Press.
Lovibond, S. (1995). Aristotelian ethics and the "enlargement of thought". In R. Heinaman (Ed.), *Aristotle and moral realism*. London: UCL Press.
Lovibond, S. (2002). *Ethical formation*. Cambridge, MA: Harvard University Press.
MacIntyre, A. (1985a). *After virtue*. London: Duckworth. 2nd ed. with postscript.
MacIntyre, A. (1985b). How psychology makes itself true – or false. In S. Koch & D. E. Leary (Eds.), *A century of psychology as science*. New York: McGraw-Hill.
MacIntyre, A. (1988). *Whose justice? Which rationality?* London: Duckworth.
MacIntyre, A. (1999a). *Dependent rational animals – Why human beings need the virtues*. London: Duckworth.
MacIntyre, A. (1999b). Social structures and their threats to moral agency. *Philosophy, 74*, 311–329.
MacIntyre, A. (2001). Virtue ethics. In L. C. Becker & C. B. Becker (Eds.), *Encyclopedia of ethics* (2nd ed.). London: Routledge.
Mackie, J. (1977). *Ethics: Inventing right and wrong*. Hammondsworth: Penguin Books.
Manicas, P. T. (2002). John Dewey and American psychology. *Journal for the Theory of Social Behaviour, 32*, 267–294.
Martin, J. (2002). *The education of John Dewey*. New York: Columbia University Press.
Martin, J., & Sugarman, J. (2001). Interpreting human kinds: Beginnings of a hermeneutic psychology. *Theory and Psychology, 11*, 193–207.
Marx, K. (1964). *Early Writings*. New York: McGraw-Hill.
Matthews, E. (2004). Merleau-Ponty's body-subject and psychiatry. *International Review of Psychiatry, 16*, 190–198.
McCrae, R. R., & Costa, P. T. (1987). Validation of the five-factor model of personality across instruments and observers. *Journal of Personality and Social Psychology, 52*, 81–90.
McDowell, J. (1981). Non-Cognitivism and Rule-Following. In S. M. Holtzman & C. M. Leich (Eds.), *Wittgenstein: To follow a rule*. London: Routledge and Kegan Paul.
McDowell, J. (1994). *Mind and world*. Cambridge, MA: Harvard University Press.
McDowell, J. (1998). *Mind, value, and reality*. Cambridge, MA: Harvard University Press.
Merleau-Ponty, M. (1945). *Phenomenology of perception*. London: Routledge. This edition published 2002.
Midgley, M. (1994). *The ethical primate: Humans, freedom and morality*. London: Routledge.
Mill, J. S. (1843). *The logic of the moral sciences*. London: Duckworth. This edition published 1987.
Miller, E. F. (1971). Hume's contribution to behavioral science. *Journal of the History of the Behavioral Sciences, 7*, 154–168.
Murdoch, I. (1997a). Metaphysics and ethics. In P. Conradi (Ed.), *Existentialists and mystics: Writings on philosophy and literature*. London: Chatto and Windus. Article first published 1957.
Murdoch, I. (1997b). The idea of perfection. In P. Conradi (Ed.), *Existentialists and mystics: Writings on philosophy and literature*. London: Chatto and Windus. Article first published 1964.
Murdoch, I. (1997c). The sovereignty of good over other concepts. In P. Conradi (Ed.), *Existentialists and mystics: Writings on philosophy and literature*. London: Penguin. First published 1967.
Nagel, T. (1986). *The view from nowhere*. Oxford: Oxford University Press.

Nagel, T. (1997). *The last word.* Oxford: Oxford University Press.

Nietzsche, F. (1889). *Twilight of the idols.* London: Penguin.

Nucci, L. P., & Turiel, E. (1978). Social Interactions and the development of social concepts in preschool children. *Child Development, 49*, 400–407.

Nussbaum, M. C. (1986). *The fragility of goodness: Luck and ethics in greek tragedy and philosophy.* Cambridge: Cambridge University Press.

Nussbaum, M. C. (1990). The discernment of perception: An Aristotelian conception of private and public rationality. In *Love's knowledge.* Oxford: Oxford University Press.

Nussbaum, M. C. (1994). *The therapy of desire: Theory and practice in Hellenistic ethics.* In Princeton, NJ: Princeton University Press.

Olssen, M. (2002). Michel foucault as "Thin" communitarian: Difference, community, democracy. *Cultural Studies – Critical Methodologies, 2*, 483–513.

Paterson, M. (2006). *Consumption and everyday life.* London: Routledge.

Pawlik, K., & Rosenzweig, M. R. (2000). *The international handbook of psychology.* London: Sage.

Piaget, J. (1932). *The moral judgment of the child.* London: Routledge and Kegan Paul. This edition published 1975.

Pinkard, T. (2007). Was pragmatism the successor to idealism? In C. Misak (Ed.), *New pragmatists.* Oxford: Oxford University Press.

Plato (1987). *The Republic.* London: Penguin.

Polkinghorne, D. (1989). Changing conversations about human science. In S. Kvale (Ed.), *Issues of validity in qualitative research.* Lund: Studentlitteratur.

Polkinghorne, D. (2000). Psychological inquiry and the pragmatic and hermeneutic traditions. *Theory and Psychology, 10*, 453–479.

Polkinghorne, D. (2004). *Practice and the human sciences.* Albany, NY: SUNY Press.

Poulsen, A. (1995). Modernity and the upgrading of psychological reflectivity. *Journal of Phenomenological Psychology, 26*, 1–20.

Putnam, H. (1973). Meaning and reference. *The Journal of Philosophy, 70*, 699–711.

Putnam, H. (2002). *The collapse of the fact/value dichotomy and other essays.* Cambridge, MA: Harvard University Press.

Quine, W. V. (1951). Two dogmas of empiricism. *Philosophical Review, 60*, 20–43.

Quine, W. V. (1969). *Ontological relativity and other essays.* New York: Columbia University Press.

Qvortrup, L. (2003). *The hypercomplex society.* New York: Peter Lang.

Reader, S. (2000). New directions in ethics: Naturalisms, reasons and virtue. *Ethical Theory and Moral Practice, 3*, 341–364.

Reath, A. (2001). Jean-Jacques Rousseau. In L. C. Becker & C. B. Becker (Eds.), *Encyclopedia of ethics* (2nd ed.). London: Routledge.

Richards, G. (1996). *Putting psychology in its place: An introduction from a critical historical perspective.* London: Routledge.

Richards, G. (2002). The psychology of psychology: A historically grounded sketch. *Theory and Psychology, 12*, 7–36.

Richardson, F. C., Fowers, B. J., & Guignon, C. B. (1999). *Re-envisioning psychology: Moral dimensions of theory and practice.* San Fransisco: Jossey-Bass.

Ricoeur, P. (1992). *Oneself as another.* Chicago: University of Chicago Press.

Rieff, P. (1965). *The triumph of the therapeutic.* Chicago: University of Chicago Press. Denne udgave 1987.

Robinson, D. N. (1984). On reasons and causes. *Annals of Theoretical Psychology, 2*, 157–163.

Robinson, D. N. (1989). *Aristotle's psychology.* New York: Columbia University Press.

Robinson, D. N. (1992). Social science and moral science. In D. N. Robinson (Ed.), *Social discourse and moral judgment.* London: Academic Press.

Robinson, D. N. (2002). *Praise and blame: Moral realism and its applications.* Princeton, NJ: Princeton University Press.

Rogers, C. (1961). *On becoming a person.* Boston: Houghton Mifflin Company.
Rogers, C. (1967a). Learning to be free. In C. Rogers & B. Stevens (Eds.), *Person to person: The problem of being human.* Moab, UT: Real People Press.
Rogers, C. (1967b). Toward a modern approach to values: The valuing process in the mature person. In C. Rogers & B. Stevens (Eds.), *Person to person: The problem of being human.* Moab, UT: Real People Press.
Rogers, C. (1970). *Becoming partners: Marriage and its alternatives.* New York: Dell.
Rosa, H. (2003). Social acceleration: Ethical and political consequences of a desynchronized high-speed society. *Constellations, 10*, 3–33.
Rose, N. (1985). *The psychological complex: Psychology, politics and society in England 1869–1939.* London: Routledge and Kegan Paul.
Rose, N. (1996a). *Inventing our selves: Psychology, power, and personhood.* Cambridge: Cambridge University Press.
Rose, N. (1996b). Power and subjectivity: Critical history and psychology. In C. F. Graumann & K. J. Gergen (Eds.), *Historical dimensions of psychological discourse.* Cambridge: Cambridge University Press.
Rose, N. (1999a). *Governing the soul: The shaping of the private self* (2nd ed.). London: Free Association Books.
Rose, N. (1999b). *Powers of freedom.* Cambridge: Cambridge University Press.
Rose, N. (2003). Neurochemical selves. *Society, 41*, 46–59.
Rose, N. (2007). *The politics of life itself: Biomedicine, power and subjectivity in the twenty-first century.* Princeton, NJ: Princeton University Press.
Rouse, J. (2002). *How scientific practices matter: Reclaiming philosophical naturalism.* Chicago: University of Chicago Press.
Rouse, J. (2007). Social practices and normativity. *Philosophy of the Social Sciences, 37*, 46–56.
Rousseau, J. J. (1762). *Émile.* London: J.M. Dent and Sons. Denne udgave 1966.
Rousseau, J.-J. (1996). *The confessions.* Ware: Wordsworth classics. First published 1782.
Ruse, M. (1991). The significance of evolution. In P. Singer (Ed.), *A companion to ethics.* Oxford: Blackwell.
Ryle, G. (1971a). The thinking of thoughts: What is ‘Le Penseur’ doing? In *Collected papers volume II: Collected essays 1929–1968.* London: Hutchinson & Co.
Ryle, G. (1971b). Thinking and reflecting. In *Collected papers volume II: Collected essays 1929–1968.* London: Hutchinson & Co.
Schatzki, T. R. (2000). Wittgenstein and the social context of an individual life. *History of the Human Sciences, 13*, 93–107.
Schwandt, T. (2000). Three epistemological stances for qualitative inquiry: Interpretivism, hermeneutics, and social constructionism. In N. K. Denzin & Y. S. Lincoln (Eds.), *Handbook of qualitative research.* London: Sage.
Searle, J. (1967). How to derive ‘ought’ from ‘is’. In P. Foot (Ed.), *Theories of ethics* (pp. 101–114). Oxford: Oxford University Press.
Seligman, M. E. P. (2002). *Authentic happiness: Using the new positive psychology to realize your potential for lasting fulfillment.* London: Nicholas Brealey Publishing.
Sellars, W. (1997). *Empiricism and the philosophy of mind.* Cambridge, MA: Harvard University Press. First published 1956.
Sennett, R. (1977). *The fall of public man.* London: Penguin. Denne udgave 2003.
Sennett, R. (1998). *The corrosion of character.* New York: Norton.
Sennett, R. (2006). *The culture of the new capitalism.* New Haven: Yale University Press.
Shotter, J. (1999). Vygotsky: The social negotiation of semiotic mediation. In P. Lloyd & C. Fernyhough (Eds.), *Lev Vygotsky: Critical assessments. Volume IV: Future directions.* London: Routledge.
Shweder, R. A. (1990). In defense of moral realism: Reply to Gabennesch. *Child Development, 61*, 2060–2067.
Shweder, R. A., Mahapatra, M., & Miller, J. G. (1990). Culture and moral development. In J. W. Stigler, R. A. Shweder, & G. Herdt (Eds.), *Cultural psychology: Essays on comparative human development.* Cambridge: Cambridge University Press.

Siegel, H. (1996). Naturalism and the abandonment of normativity. In W. O'Donohue & R. F. Kitchener (Eds.), *The philosophy of psychology*. London: Sage.
Skinner, B. F. (1965). *Science and human behavior*. New York: The Free Press. First published 1953.
Skinner, B. F. (1971). *Har vi råd til frihed?* Haslev: Gyldendal.
Smith, L. (2006). Norms in human development. In L. Smith (Ed.), *Norms in human development*. Cambridge: Cambridge University Press.
Smith, N. H. (2002). *Charles Taylor: Meaning, morals and modernity*. Cambridge: Polity Press.
Smith, R. (1997). *The Norton history of the human sciences*. New York: W.W. Norton & Co.
Stengers, I. (2000). *The invention of modern science*. Minneapolis: University of Minnesota Press.
Stenner, P. (1998). Heidegger and the subject. *Theory and Psychology, 8*, 59–77.
Stern, D. (2000). Practices, practical holism, and background practices. In M. Wrathall & J. Malpas (Eds.), *Heidegger, coping, and cognitive science – Essays in honor of Hubert L. Dreyfus, volume 2*. Cambridge, MA: MIT.
Still, A., & Good, J. (1998). The ontology of mutualism. *Ecological Psychology, 10*, 39–63.
Strickland, B. (2001). *The gale encyclopedia of psychology* (2nd ed.). Detroit: Gale Group.
Susman, W. I. (1979). "Personality" and the making of twentieth century culture. In J. Higham & P. K. Conkin (Eds.), *New directions in American intellectual history*. Baltimore: The Johns Hopkins University Press.
Szasz, T. (1961). *The myth of mental illness: Foundations of a theory of personal conduct*. New York: HarperCollins.
Szasz, T. (1994). *Cruel compassion: The psychiatric control of society's unwanted*. Syracyse: Syracuse University Press.
Taylor, C. (1971). Interpretation and the sciences of man. *Review of Metaphysics, 25*, 3–51.
Taylor, C. (1975). *Hegel*. Cambridge: Cambridge University Press.
Taylor, C. (1981). Understanding and explanation in the *Geisteswissenschaften*. In S. M. Holtzman & C. M. Leich (Eds.), *Wittgenstein: To follow a rule*. London: Routledge and Kegan Paul.
Taylor, C. (1984). Foucault on freedom and truth. *Political Theory, 12*, 152–183.
Taylor, C. (1985a). Atomism. In *Philosophy and the human sciences: Philosophical papers 2*. Cambridge: Cambridge University Press.
Taylor, C. (1985b). Self-interpreting animals. In *Human agency and language: Philosophical papers 1*. Cambridge: Cambridge University Press.
Taylor, C. (1985c). Social theory as practice. In *Philosophy and the human sciences: Philosophical papers 2*. Cambridge: Cambridge University Press.
Taylor, C. (1985d). What is human agency? In *Human agency and language: Philosophical papers 1*. Cambridge: Cambridge University Press.
Taylor, C. (1988). The moral topography of the self. In S. B. Messer, L. A. Sass, & R. L. Woolfolk (Eds.), *Hermeneutics and psychological theory*. New Brunswick: Rutgers University Press.
Taylor, C. (1989). *Sources of the self*. Cambridge: Cambridge University Press.
Taylor, C. (1991). *The ethics of authenticity*. Cambridge, MA: Harvard University Press.
Taylor, C. (1993). Engaged agency and background in Heidegger. In C. B. Guignon (Ed.), *The Cambridge companion to heidegger*. Cambridge: Cambridge University Press.
Taylor, C. (1994). Justice After Virtue. In J. Horton & S. Mendus (Eds.), *After macintyre: Critical perspectives on the work of Alasdair MacIntyre*. Cambridge: Polity Press.
Taylor, C. (1995a). Explanation and practical reason. In *Philosophical Arguments*. Cambridge, MA: Harvard University Press.
Taylor, C. (1995b). Lichtung or lebensform: Parallels between heidegger and wittgenstein. In *Philosophical arguments*. Cambridge, MA: Harvard University Press.
Taylor, C. (1995c). Overcoming Epistemology. In *Philosophical arguments*. Cambridge, MA: Harvard University Press.
Taylor, C. (1996). Iris Murdoch and moral philosophy. In M. Antonaccio & W. Schweiker (Eds.), *Iris Murdoch and the search for human goodness*. Chicago: The University of Chicago Press.
Taylor, C. (1999). Two theories of modernity. *Public Culture, 11*, 153–174.

Taylor, C. (2000). What's wrong with foundationalism?: Knowledge, agency, and world. In M. Wrathall & J. Malpas (Eds.), *Heidegger, coping, and cognitive science: Essays in Honor of Hubert L. Dreyfus volume 2*. Cambridge, MA: The MIT Press.

Taylor, C. (2002). Modern social imaginaries. *Public Culture, 14*, 91–124.

Taylor, C. (2004). *Modern social imaginaries*. Durham, NC: Duke University Press.

Taylor, C. (2007). *A secular age*. Cambridge, MA: Harvard University Press.

Teehan, J. (1996). In defense of a naturalism. *The Journal of Speculative Philosophy, 10*, 79–91.

Tooby, J., & Cosmides, L. (1992). The psychological foundations of culture. In J. H. Barkow, L. Cosmides, & J. Tooby (Eds.), *The adapted mind*. Oxford: Oxford University Press.

Toulmin, S. (1990). *Cosmopolis: The hidden agenda of modernity*. Chicago: University of Chicago Press.

Toulmin, S. (2001). *Return to reason*. Cambridge, MA: Harvard University Press.

Turiel, E. (1983). *The development of social knowledge: Morality and convention*. Cambridge: Cambridge University Press.

Turiel, E. (1998). The development of morality. In W. Damon (Ed.), *Handbook of child psychology, vol. 3: Social, emotional and personality development* (5th ed.). New York: Wiley.

Turiel, E., & Wainryb, C. (2000). Social life in cultures: Judgments, conflict, and subversion. *Child Development, 71*, 250–256.

Urban, H. B. (1983). Phenomenological-humanistic approaches. In M. Hersen, A. E. Kazdin, & A. S. Bellack (Eds.), *The clinical psychology handbook*. New York: Pergamon Press.

van Alstyne, L. (1998). Aristotle's alleged ethical obscurantism. *Philosophy, 73*, 429–452.

van Drunen, P., van Strien, P. J., & Haas, E. (2004). Work and organization. In J. Jansz & P. van Drunen (Eds.), *A social history of psychology* (pp. 129–164). Oxford: Blackwell.

Viner, R. (1999). Putting stress in life: Hans selye and the making of stress theory. *Social Studies of Science, 29*, 391–410.

Vitz, P. C. (1994). *Psychology as religion: The cult of self-worship* (2nd ed.). Grand Rapids: William B. Eerdmans Publishing Company.

Vygotsky, L. S. (1978). *Mind in society: The Development of higher psychological processes*. Cambridge, MA: Harvard University Press.

Ward, S. C. (2002). *Modernizing the mind: Psychological knowledge and the remaking of society*. Westport: Praeger.

Watson, J. B. (1914). *Behaviorism: An introduction to comparative psychology*. New York: Holt.

Williams, B. (1985). *Ethics and the limits of philosophy*. London: Fontana.

Williams, R. (1978). *Marxism and literature*. Oxford: Oxford University Press.

Williams, R. (1983). *Keywords: A vocabulary of culture and society*. London: Fontana. First published 1976.

Williams, R. N., & Beyers, M. S. (2001). Personalism, social constructionism and the foundation of the ethical. *Theory and Psychology, 11*, 119–134.

Williams, R. N., & Gantt, E. E. (1998). Intimacy and heteronomy: On grounding psychology in the ethical. *Theory and Psychology, 8*, 253–267.

Williams, R. N., & Gantt, E. E. (2002). Pursuing psychology as science of the ethical: Contributions of the work of Emmanuel Levinas. In E. E. Gantt & R. N. Williams (Eds.), *Psychology for the other: Levinas, ethics and the practice of psychology* (pp. 1–31). Pittsburgh: Duquesne University Press.

Wilson, E. O. (1975). *Sociobiology: The new synthesis*. Cambridge, MA: Harvard University Press.

Winch, P. (1963). *The idea of a social science and its relation to philosophy*. London: Routledge and Kegan Paul. First published 1958.

Wittel, A. (2001). Toward a network sociality. *Theory, Culture and Society, 18*, 51–76.

Wittgenstein, L. (1953). *Philosophical investigations*. Oxford: Basil Blackwell.

Wokler, R. (2001). *Rousseau*. Oxford: Oxford University Press.

Wright, R. (1994). *The moral animal: Why we are the way we are: Evolutionary psychology and everyday life*. New York: Pantheon Books.

Yanchar, S. C., & Hill, J. R. (2003). What is psychology about? Toward an explicit ontology. *Journal of Humanistic Psychology, 43*, 11–32.

Index

MIX
Papier aus verantwortungsvollen Quellen
Paper from responsible sources
FSC® C105338

If you have any concerns about our products,
you can contact us on
ProductSafety@springernature.com

In case Publisher is established outside the EU,
the EU authorized representative is:
Springer Nature Customer Service Center GmbH
Europaplatz 3, 69115 Heidelberg, Germany

Printed by Libri Plureos GmbH
in Hamburg, Germany